FROM MOBILITY TO ACCESSIBILITY

FROM MOBILITY TO ACCESSIBILITY

Transforming Urban Transportation and Land-Use Planning

Jonathan Levine
Joe Grengs
Louis A. Merlin

CORNELL UNIVERSITY PRESS ITHACA AND LONDON

First published 2019 by Cornell University Press

Library of Congress Cataloging-in-Publication Data

Names: Levine, Jonathan (Jonathan C.), author. | Grengs, Joe, 1962– author. | Merlin, Louis A., 1972– author.
Title: From mobility to accessibility : transforming urban transportation and land-use planning / Jonathan Levine, Joe Grengs, Louis A. Merlin.
Description: Ithaca [New York] : Cornell University Press, 2019. | Includes bibliographical references and index.
Identifiers: LCCN 2019015736 (print) | LCCN 2019018066 (ebook) | ISBN 9781501716102 (pdf) | ISBN 9781501716096 (epub/mobi) | ISBN 9781501716072 | ISBN 9781501716072 (cloth) | ISBN 9781501716089 (pbk.)
Subjects: LCSH: Urban transportation—Planning. | Urban transportation—United States—Planning. | Land use, Urban—Planning. | Land use, Urban—United States—Planning. | City planning. | City planning—United States.
Classification: LCC HE305 (ebook) | LCC HE305 .L48 2019 (print) | DDC 388.4068/4—dc23
LC record available at https://lccn.loc.gov/2019015736

Dedicated in loving memory

Rose Aronowitz Levine
Hillel Benjamin Levine
Judith Langer Merlin

Contents

Acknowledgments

We are grateful to the many individuals and organizations that helped make this book possible. Indispensable institutional support came from the Rockefeller Foundation Bellagio Center, the Environmental Protection Agency, the Graham Sustainability Institute, the Dow Sustainability Postdoctoral Fellowship, the Federal Highway Administration, the NEXTRANS Center, and the Taubman College of Architecture and Urban Planning at the University of Michigan. Data were generously contributed by the City of Ann Arbor, the Accessibility Observatory at the University of Minnesota, the Southeast Michigan Council of Governments, the Alamo Area Metropolitan Planning Organization, and dozens of other MPOs around the United States.

Research findings in chapter 3 were previously published in Jonathan Levine, Louis Merlin, and Joe Grengs, "Project-Level Accessibility Analysis for Land-Use Planning," *Transport Policy* 53 (2017): 107–19, and in Louis A. Merlin, Jonathan Levine, and Joe Grengs, "Accessibility Analysis for Transportation Projects and Plans," *Transport Policy* 69 (2018): 35–48. Research findings in chapter 4 were previously published in Jonathan Levine, Joe Grengs, Qingyun Shen, and Qing Shen, "Does Accessibility Require Density or Speed?: A Comparison of Fast versus Close in Getting Where You Want to Go in U.S. Metropolitan Regions," *Journal of the American Planning Association* 78, no. 2 (2012): 157–72. All used by permission.

Qing Shen and Qingyun Shen were our partners in much of the research reported in chapters 4 and 7, and Qingyun provided brilliant assistance throughout much of the empirical research of the book. We benefited from superb research assistance as well from Matan Singer, Alexandra Markiewicz, Doug Koloszvari, Jason Enos, Patrick Dieter, Arthur Prokosch, Wenting Chen, and Chao Liu. Zachary Graham, from the Alamo Area Metropolitan Planning Organization, provided assistance with data and modeling. Guillaume Barreau programmed a website to render methods in chapter 3 accessible to planners in practice. Prashanth Raju enhanced the consistency and clarity of the graphics. Two anonymous reviewers and our editor, Michael McGandy, offered constructive feedback that greatly improved the draft. Julie Steiff similarly improved the text with her professional editing, and we benefited from comments on previous drafts from colleagues, students, and family including

Jacob Yan, Matan Singer, Adam Levine, Rob Goodspeed, Adam Millard-Ball, Deena Levine, and Michael Lipsett.

Finally, our gratitude goes to Noga Morag-Levine, Suzanne Copsey, and Ki-Wing Merlin for their loving support throughout the process of creating this book.

Abbreviations

BRT	Bus Rapid Transit
CEQA	California Environmental Quality Act
EIR	Environmental Impact Report
EJ	Environmental Justice
FHWA	Federal Highway Administration
FTA	Federal Transit Administration
GTFS	General Transit Feed Specification
LOS	Level of Service
LRT	Light Rail Transit
MAP-21	Moving Ahead for Progress in the 21st Century Act
MPO	Metropolitan Planning Organization
MRCOG	Mid-Region Council of Governments
MSA	Metropolitan Statistical Area
NAICS	North American Industry Classification System
NEPA	National Environmental Policy Act
NHTS	National Household Travel Survey
NMDOT	New Mexico Department of Transportation
NTD	National Transit Database
SMART	System Management and Allocation of Resources for Transportation
TAZ	Traffic Analysis Zone
TIGER	Transportation Investment Generating Economic Recovery
TIP	Transportation Improvement Plan
VKT	Vehicle-Kilometers Traveled
VRM	Vehicle Revenue Miles

FROM MOBILITY TO ACCESSIBILITY

Introduction

THE ACCESSIBILITY SHIFT

Imagine that you have a list of tasks to accomplish that requires you to travel to a number of destinations today: work, shopping, day care, and maybe a restaurant or cinema. Unlike a typical day, today you have the freedom to choose between two metropolitan areas to meet these ordinary needs. While both regions host the full range of destinations, the first region boasts more rapid surface travel—less congestion, faster highways, and more rapid transit—than the second. The second has slower travel, but your destinations are closer to your starting point and to one another, such that you can accomplish your task list with less investment of time and money. Based solely on your ability to travel quickly, region number one would seem to be the better choice. But knowledge about travel speeds would not give you a complete picture of your ability to meet your needs in that region; the relative locations of your destinations would matter as well. If you choose the second region because its closer destinations more than offset the slower travel it supports, you are selecting the higher-accessibility option.

Ironically, you are choosing the region that current transportation-planning methods identify as problematic.[1] Since the mid-twentieth century, transportation outcomes, together with the transportation dimension of land use, have largely been planned and evaluated within a framework based not in accessibility, but in mobility. Faster movement, particularly by cars, is treated as the central goal, without attention to accessibility as an outcome. In this book we advocate a shift from mobility to accessibility as the basis for transportation and land-use planning, arguing that such a shift (a) is compelled by the very purpose of transportation; (b) would be transformative to the practices of both transportation planning and land-use planning; but (c) is impeded by numerous misconceptions. The misconceptions constitute impediments to the mobility-to-accessibility shift, termed here "the accessibility shift," impediments we seek to help overcome.

Accessibility and Mobility

The concepts of *accessibility* and *mobility* are readily distinguished by considering what would count as an improvement in each.[2] An improvement in mobility is an increase in the *territory* that can be reached for a given investment of time and money, while an improvement in accessibility is an increase in the *destinations* that can be reached for a given investment of time and money. To be more precise, destinations vary in their value; most people would appreciate access to a supermarket more than to a convenience store, for example. So an accessibility improvement is more precisely defined as an increase in the *value of destinations* that can be reached for a given investment of time and money. Physical accessibility, the focus of this book, refers to the potential for traveling to actual destinations, but the concept can be broadened to accessibility more generally by including remote electronic interaction with virtual destinations and delivery of goods and services to one's location. While physical accessibility varies widely by travel mode, it is equally relevant to auto drivers, public-transport riders, pedestrians, and cyclists; it is the principal service that people seek from transportation regardless of the means they choose. And though improving accessibility may help further planning goals including social equity, environmental performance, and multimodalism—and may be motivated in part by the struggle to achieve them—accessibility is not equivalent to those goals. Accessibility has, in some cases, become a shorthand for a long wish list of desirable transportation outcomes; we reject this expansive view in favor of the more limited, and more workable, definition given here.

Reliance on mobility as a guiding planning principle is evidenced in current policy and in the physical form of the built environment in metropolitan areas in the United States and many other countries around the world. When evaluating the performance of a transportation system, the fundamental criterion for success has long been faster vehicle-operating speed.[3] Common indicators of this include delay per capita, dollars wasted while waiting in traffic, and highway level-of-service.[4] These mobility metrics are not simply after-the-fact assessments but are also used proactively to guide policy toward transportation investment and land-use regulation.[5] On the ground, this often translates into a singular focus on fast and wide highways and land-use regulations that strive to forestall traffic congestion by limiting allowable development heights or densities, a combination of policies that can increase metropolitan travel distances by inducing an increasing spread of origins and destinations.[6]

This focus on mobility stands in contrast to a cornerstone of modern transportation planning: the notion that the demand for transportation is largely derived from the demand to reach destinations.[7] Economists refer to a *derived demand* as a demand for an intermediate good or service that occurs as a result of the demand for a final good or service. Examples include fertilizer purchased by farmers in order to grow crops or electrical power consumed in order to attain desired services such as illumination or computing power. Evaluating farmers' productivity with metrics of fertilizer consumption rather than crop production would be an obvious error that, if used to guide decision making, would surely lead to bad policy. Transportation planning is largely built on a similar error, because the derived-demand view, a consensus across the transportation professions, implies that mobility is primarily a means in transportation rather than an end.

If the goal that drives current transportation planning—mobility—differs from the service that people seek from transportation—accessibility—the planning process would tend to overprovide the former and underprovide the latter. The transportation and land-use systems would require a high degree of person and vehicle movement to overcome the great distances between origins and destinations. A shift to an accessibility logic would add proximity to the lexicon and tool kit of the transportation planner. Where mobility-based transportation planning has been largely indifferent to the distance between origins and destinations, a shift toward accessibility would seek to create gains for current and future residents through mobility and proximity in combination. Aligning the logic of transportation planning with the core purpose of transportation is about ensuring that people get more of what they want out of the system: access to destinations.

And the observation that transportation demand is derived is in fact a function of people and what they want. Transportation demand is not derived as a matter of definition or axiom; if people were observed to behave as if they were usually seeking movement for its own sake when they travel, the derived-demand conclusion would be unsupported. While some travel is clearly for the sake of movement, the empirical evidence strongly supports the idea that, for a large majority of trips, time spent in travel is a cost that people bear in order to reap the benefits available at their destinations (see chapter 1).

Derived Demand and the Accessibility Shift

The derived nature of most transportation demand (which we will refer to as the derived-demand framework) is not at all controversial. Quite the opposite: this fundamental concept is embedded throughout mainstream transportation-planning methods.[8] For example, travel-demand models rely on the arrangement of land-use patterns across a region to forecast travel flows; this capacity to forecast travel largely on the basis of the spatial arrangement of origins and destinations is a direct outgrowth of the derived-demand framework.[9] If transportation were undertaken simply for the purpose of movement, then the ease with which destinations could be reached would not be a useful predictor of trip making in travel-demand models.

Derived demand is also at the core of traditional cost-benefit evaluation methods for transportation planning. Consider, for example, a bypass route that allows travelers to reach their destination more quickly. Cost-benefit analysis considers the travelers' time savings to be the principal relevant benefit of the infrastructure investment. After the investment, the travelers "consume" less mobility because they can now take a more direct route to their destination. If movement itself were the service travelers desired, switching to a shorter highway route would represent people's consuming less of what they want; it would paradoxically be considered a harm rather than a benefit. The derived-demand framework avoids this dilemma by viewing mobility correctly as an intermediate service that is consumed for the sake of reaching one's destination; in this way derived demand is implicit in all realms in transportation planning and policy.

But while the fields of transportation planning, engineering, and economics are all based on this destination-oriented view of transportation, there is an implication that they have too rarely confronted. If the demand for travel is indeed mostly derived, then assessments of the quality or speed of movement, when performed in isolation, cannot reveal the extent to which people's transportation needs are being met. Such an assessment would need to change to one based on people's capacity to reach destinations, a function of both the speed of their travel and destination proximity, not just their ease of movement. In this way, the accessibility shift, a definition of what constitutes success in transportation, is compelled as a matter of logic by the derived nature of transportation demand, which is a matter of empirical human experience.

This accessibility shift would align transportation and land-use planning with thinking in other professions. For example, site selection for retail facilities has for decades been driven by accessibility considerations.[10] The real-estate truism that the components of value are "location, location, location" is, in large part, an accessibility-related assertion, and measured accessibility has in fact long been a key component of land-value models.[11] Real estate agents know that people frequently express appreciation for their home and neighborhood on the basis of the destinations reachable from them; rarely do they express the appreciation based on the travel speeds on the roadways serving their locales. The accessibility shift does not require transportation and land-use planners to convince others to begin to act in an accessibility-aware fashion; it does, however, demand reform of the principles and norms of the profession itself.

Neither does the shift represent new governmental intervention into markets. The practice of transportation and land-use planning, with its massive public investment in transportation infrastructure, detailed and prescriptive regulation of land use through municipal zoning and subdivision controls, and reliance on tools including taxation, eminent domain, and subsidy, is inherently an exercise in governmental intervention into markets.[12] The accessibility shift seeks to align that ubiquitous intervention with transportation's purpose.

Grounding the accessibility shift in the derived-demand framework contrasts with some traditions in accessibility-related thinking. The primary benefit of the accessibility shift is increasing people's capacity to reach destinations, yet accessibility is frequently seen first and foremost as a means to yield other, ancillary benefits aligned with current policy imperatives: reductions in vehicle kilometers traveled (VKT) and attendant pollution and greenhouse-gas emissions; improvements in the equity of the transportation and land-use system; increases in active transportation including walking and cycling; containment of metropolitan sprawl; or even reductions in roadway congestion.[13] All these outcomes are desirable, and many of them may plausibly be aided by a shift from mobility to accessibility as the central framework of transportation and land-use planning. They may even be the prime motivators for many who might seek a shift to accessibility-based planning.

But evidence of accessibility's ancillary benefits, however welcome, ought not to be the desirability test of, or necessary precondition for, the accessibility shift. To see this, consider the universe of mobility-increasing actions. Some of them will increase accessibility, while others will leave accessibility unchanged or even degrade it. If the purpose of most travel is reaching destinations, the latter two sets of actions fail to further that purpose (and the last undermines it). Hence the derived-demand framework implies rejection of a mobility criterion for transportation success when considered in isolation. By contrast, consider the universe of accessibility-increasing actions. Some will improve mobility, while others will leave it unchanged or even degrade it. All of them, however, would further transportation's purpose by improving people's ability to reach destinations. This is sufficient to establish their transportation benefit; any benefits in other realms would be a bonus. In this way, accessibility is much more than one desirable framework, a "politically attractive concept," as Ilan Salomon and Patricia Mokhtarian assert, or Gerald Houseman's "handy term to use in discussions of mobility . . . quite popular among . . . environmentally conscious writers."[14] The derived-demand framework implies that accessibility is an inherent benefit apart from any ancillary benefits it might also provide. This idea is the derived-demand rationale for the accessibility shift.[15]

The inherent, rather than instrumental, view of the accessibility shift faces a significant obstacle. Concerns about environmental performance, human health, or congestion are institutionalized through government agencies at various levels, nongovernmental organizations, and private-sector mobilization. By contrast, there are few integrated transportation/land-use agencies, or "departments of accessibility," that would be well positioned to champion the accessibility shift. For this reason, the temptation to condition the shift on already-institutionalized concerns is great. But that linkage comes at a cost: while evidence of these ancillary gains can add political urgency to the accessibility shift, such conditioning implies that scientific uncertainty regarding accessibility's purported benefits in areas such as emissions reductions or health improvements—an ever-present feature of large societal questions such as these—would be an argument against the shift. Yet rejecting the accessibility shift does not amount to a neutral default, but rather a reversion to mobility-based planning and evaluation.

The derived-demand rationale may be particularly important given the time scales over which behavioral shifts operate. For example, consider the travel-behavior impacts of developing an individual site in a walkable or transit-friendly fashion when the regional situation for that neighborhood is auto-oriented. Any VKT-reduction effects are likely to be slight because so much of the travel from the neighborhood is to region-wide destinations and will continue to be by car.[16] Policy reform in this realm, however, aspires not merely to islands of walkable urbanism within a sea of sprawl, but to overall transformation in metropolitan-development trends. Hoped-for reductions in driving are thus much greater than those that isolated pockets could ever produce (even on a per-capita basis), yet proof of these greater benefits will always be uncertain in the near term since cities change slowly. Accessibility is the immediate benefit that can be pursued as policy reformers seek the more uncertain payoff of long-run driving reductions based on shifts in patterns of development.

Our primary argument—that the derived-demand framework compels an accessibility perspective and establishes accessibility as the appropriate evaluative framework for transportation planning—has not always been evident. For example, derived demand has been viewed as impeding rather than compelling the accessibility shift. "Perhaps a major reason that the analysis of accessibility has not been frequently proposed as a basic component of a social report, is the common notion that movement is a 'derived' demand. . . . It could be argued that because transportation and physical accessibility constitute means to achieving ends such as income, health, and education, accessibility need not be singled out as a component of a set of indicators," wrote Martin Wachs and Gordon Kumagai in 1973.[17] In other cases, authors cite the derived-demand watchword but fail to acknowledge its inconsistency with a mobility-based definition of the purposes of transportation, as Jean-Paul Rodrigue wrote in 2009: "What are the differences between a Boeing 747, an oil tanker, a car and a bicycle? Extensive indeed, but they each share the common goal of fulfilling a derived transport demand, and they thus all fill the purpose of supporting mobility."[18]

Some authors go further, portraying accessibility as a means to mobility rather than the reverse, despite explicit acknowledgment of derived demand. "Unsuccessful solutions to this problem [congestion] and the wide recognition of the effects of accessibility has [*sic*] resulted in the idea that by improving accessibility, reductions in congestion . . . could be attained. . . . In this approach travel is considered a derived demand, which implies that the decision to travel or not originates from the need to go to a particular destination," wrote Irene Casas in 2003.[19]

A Distinct Relationship between Accessibility and Mobility

Traditionally, transportation planning and engineering have defined themselves in mobility terms; transportation investments or policy interventions including highway-capacity expansion, public-transport deployment, and even land-use planning are evaluated in terms of their capacity to promote mobility, which is often interpreted as prevention or alleviation of highway congestion.[20] This focus on mobility needs to be altered in light of the derived-demand framework. Mobility would become one means to achieve accessibility; other possible pathways would be proximity and connectivity (figure I.1). Connectivity here refers to the delivery of goods and services to one's location, whether virtually (e.g., internet downloads) or physically (e.g., package delivery, traditional utilities). But mobility and proximity exist in tension with each other. Compact places

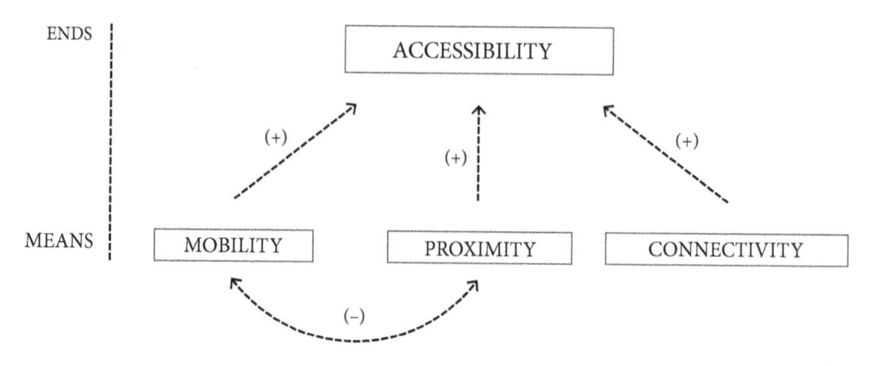

FIGURE I.1 Relationships among mobility, proximity, connectivity, and accessibility in a revised hierarchy of transportation goals

with many origins and destinations near one another tend to be locations where surface transportation is slow; conversely, areas of rapid surface travel tend to be low-density areas where origins and destinations are more dispersed.

The relationship between mobility and accessibility runs through urban form, or the patterns of origins and destinations across a region. It is not obvious by definition which urban forms offer higher accessibility: areas of rapid surface travel and low proximity, or areas offering high proximity of origins and destinations but slower travel. The accessibility impacts of moving from an area with slow travel and nearby destinations to one with faster travel and spread-out land uses would be a function of the net effect of speed and distance change between the regions. For example, if the region with widely spread origins and destinations had travel speeds for a given mode that more than made up for its great distances, it would be the higher-accessibility region of the two. For this reason, as an evaluative framework, accessibility does not inherently favor any particular urban form, but rather seeks transportation and land-use systems that support people's ability to meet their needs through interaction with their destinations.

Improved mobility does not guarantee improved accessibility and can even be associated with degraded accessibility over the long term because of land-use change. For this reason, current evaluations of transportation outcomes based in mobility alone suffer from a distinct logical flaw. If the purpose of transportation is not movement but access, then a mobility improvement that is associated with degraded accessibility would leave people with less time and money with which to interact with their destinations, and for that reason could not be viewed as a transportation-system improvement. While actual impacts would be a matter for empirical assessment (see chapter 4), the possibility of mobility improvements that fail to improve, or even degrade, accessibility argue for a reform of the definition of transportation success.

Accessibility as Potential for Interaction

Accessibility is an intangible potential for interaction as opposed to concrete revealed travel behavior. It may be difficult to believe that improvements in the potential for interaction that do not materialize as changes in trip making can be a meaningful indicator of anything, least of all the quality of a transportation system, but accessibility, rather than revealed travel behavior, is the meaningful indicator of these benefits. Consider, for example, average commuting time—a measure of revealed travel behavior, rather than accessibility—as a candidate indicator. A low average commuting time might or might not indicate great benefits from a location, since vastly different areas will demonstrate similar values; for example, at 18.9 minutes, downtown Washington, DC, and rural Burdett, Kansas (population 247), have identical average commute times.[21] These short average times are the products of a concentration of nearby job locations in the former case, and an absence of commuting opportunities in the latter; transportation-system benefits accruing to the two would not be remotely comparable. Another example is found in contrasting travel behavior by gender. Females in two-breadwinner households commute 33 percent less time on average than their male counterparts; one would not infer from that statistic that women receive greater transportation-system benefits than men.[22]

The benefit captured by accessibility, as opposed to revealed travel behavior, may be illustrated with a hypothetical scenario. Consider, for example, two lovers of Italian food. Both travel once a week to their chosen restaurant. But while the first restaurant lover has ten such restaurants reachable within a given time, the second has only one. Though their travel is similar, the benefit each receives from destinations in the environment differs greatly, across at least five dimensions.

Choice

On any given evening, the first restaurant lover has more choices than the second. Even if those choices do not translate into week-to-week variation in the restaurant selected, they change the meaning of the experience for the first restaurant lover. Choosing a destination from ten possible choices is more valuable than traveling to that same destination due to the lack of options.

Variety

If the presence of choice in fact materializes as week-to-week variations in restaurant selection, the first restaurant lover is experiencing the value of variety, and the second is not. Avoiding repetitiveness is, for many people, a component of the utility that their environment offers them.

Flexibility

People like the first restaurant lover who can reach multiple destinations enjoy the assurance that, should their circumstances, needs, or tastes change, they will be able to alter their destinations in response. This capacity is particularly important with regard to workplace accessibility. Households choose residential locations while taking into account accessibility to employment concentrations and not merely travel time to current job locations. The strategy is a logical response to the potential for future shifts in life circumstances such as a change in job or household composition.

Competition

The person who has ten restaurants from which to choose will probably be eating tastier food during a given evening than the person who has only one. In a similar fashion, workers who can travel to many potential employers tend to be in a position to strike a better deal on wages and working conditions.

Spillovers

The restaurant lover with access to ten restaurants will, in all likelihood, have ancillary access to neighbors with similar tastes. The potential for interaction with those neighbors provides spillover benefits beyond simple food enjoyment. Spillover benefits are of particular importance in the economy of metropolitan regions, in which the capacity of many people and firms to interact, or agglomeration economies, underpins high levels of productivity. Relevant interactions are not limited to travel to conventional destinations, but include the chance encounter, the overheard conversation, the ready availability of role models, and more. While the empirical analyses presented in this book measure accessibility on the basis of well-defined destinations, spillover benefits extend to the potential for planned and unplanned interactions of all types.

For a region, greater potential for interaction is preferable over less for a given level of externalities, such as crowding or pollution, which can accompany increasing accessibility. There may be instances in which individuals find little benefit in added accessibility, as with the addition of the second Starbucks in the neighborhood or the second Walmart within driving range. For that matter, access to a supermarket may be of little value to the household that prefers health food stores. While increases in the value of accessibility to the individual may well approach the vanishing point, the region as a whole under these scenarios benefits from increased accessibility in the form of potential interactions with coffee shops, discount retailers, or food stores.

Obstacles to the Accessibility Shift

The accessibility concept was brought into the urban planning field by Robert Murray Haig and later by Walter Hansen and has been the subject of, or a tool for, a virtual cottage industry of academic researchers ever since.[23] But in practice, accessibility metrics are used sparingly and have not supplanted their mobility counterparts in transportation evaluation. While practitioners have tentatively begun to use the language of accessibility, its application remains largely at the conceptual and visioning levels.

The transition to accessibility-based planning will not be complete—in fact, it cannot even really begin—until accessibility-based performance measures supplant metrics such as highway level of service in evaluating transportation success and guiding transportation and land-use decision making. Yet the transfer of accessibility-based evaluation to planning and engineering practice and decision making has been impeded by a family of interrelated obstacles.[24] Many of these are rooted in external political factors, including the scale mismatch between municipal land-use planning and regional transportation planning, entrenched interests of people and institutions that benefit from mobility-based transportation planning, and traditions of localist control.

Other obstacles are more within the purview of transportation and land-use planners and researchers themselves; these are rooted in conceptual and analytical challenges associated with accessibility. The most obvious is the relative complexity of accessibility metrics, especially when compared with their mobility counterparts. Measures like traffic speeds, highway level of service, and time lost in congestion are directly observable and unidimensional constructs that have the added advantage of being interpretable in easy-to-grasp terms like speed or hours. By contrast, accessibility is concerned with the people and places that transportation systems serve. Accessibility metrics therefore incorporate, at a minimum, attributes of the transportation and land-use systems.[25] This tends to render them more difficult to gauge and to interpret than many of those based in mobility.

But the obstacle of analytic complexity is diminishing. With increasing sophistication and distribution of tools for spatial analysis, together with spatial datasets, planners' technical capacity to evaluate accessibility can lower this barrier and potentially eliminate it altogether. Moreover, the needed inputs for accessibility metrics are regularly developed as part of travel-demand modeling at the regional level, a mainstay of transportation planning in many countries around the world. Thus, the technology, data, and methods for a significant shift to accessibility-based transportation planning are readily available.

On a broader level, the accessibility shift is impeded by institutional factors. For example, mobility models and metrics persist in their central role in transportation and land-use policy in part because they represent professional norms in the transportation fields. Highway level of service is a central metric of transportation engineering practice and is codified and updated regularly through the Transportation Research Board's *Highway Capacity Manual*.[26] Land-use decisions are constrained by adequate public facilities ordinances or statewide concurrency requirements mandating traffic-impact analysis, a technique whose methods are codified by the Institute of Transportation Engineers.[27] Practitioners bound to uphold their professional norms by legal or ethical constraints, or simple convention, would find themselves hard-pressed to deviate from established standards such as these.

While acknowledging these obstacles, we focus on conceptual impediments that stem from beliefs about what accessibility is or ought to be, or what would count as evidence in favor of the accessibility shift. Since these barriers are largely internal to urban- and transportation-planning discourse, they are perhaps more readily overcome than the external obstacles.

Travel-Cost Minimization Fallacy

The validity of the derived-demand concept and the accessibility paradigm is sometimes presumed to require that people choose residential locations, routes, modes, or destinations so as to minimize transportation costs (see chapter 1). Yet nothing in derived demand or the accessibility concept requires that transportation costs trump all other considerations related to travel or residential location choice. The fact that travelers may choose more travel than the bare minimum because of concerns about the quality of opportunities at their destinations, or the desired characteristics of their homes and neighborhoods, is evidence that they weigh the impulse to reduce travel costs against other attributes of the trip. It is neither evidence that travel is undertaken for its own sake nor evidence against the notion that access to destinations is the service that most travelers seek from their transportation systems.

Accessibility-as-Urbanism Fallacy

"Accessibility" sometimes stands for a set of investments and policies that promote pedestrianism, cycling, transit use, and compact development.[28] Its counterpart, "mobility," stands in this formulation for priority given to vehicular movement, highway investments, and auto-oriented development patterns. But an investment in automotive infrastructure can be accessibility increasing, and dense development can potentially degrade accessibility if it slows movement down too much. Moreover, even if one assumes that most urbanist policies are accessibility increasing, it does not follow that most accessibility-increasing policies are urbanist. Ironically, an equation by definition between accessibility and urbanist policies has been promoted by some accessibility advocates themselves, and in doing so, they have unwittingly enabled a stereotype of accessibility as relevant to a niche market rather than the theoretically sound alternative to mobility-based transportation planning (see chapter 2). As a practical matter, denser regions in the United States tend in fact to offer more accessibility than their lower-density counterparts (see chapter 4), but this is a matter of empirical observation rather than definition.

Positive vs. Normative Accessibility

The view that accessibility promotion is inherently valuable as a normative goal—rather than just a predictive variable—has at times been controversial. Social scientists in general are more comfortable with analysis and prediction than with prescription, and scholars have been more inclined to use measured accessibility as a predictive variable in complex models of land use, real-estate prices, or travel behavior than to transform planning practice (see chapter 2).[29] Researchers develop increasingly sophisticated accessibility metrics based on ever more fine-grained spatial data, but their focus on modeling exercises implicitly assumes either that others will make the link from the laboratory to practice or that measured accessibility has little to do with the professional practice of planning or transportation engineering. By contrast, we seek to model accessibility analyses that are designed to support planning and policy by explicitly framing policy-relevant variables in applied models where accessibility is the dependent variable or outcome of interest, rather than an independent variable (see chapters 3–8).

Scenario- versus Project-Based Accessibility

Embedding accessibility as a normative planning goal entails an additional analytical shift. Formal accessibility analyses began to appear broadly in professional planning practice in the first two decades of the twenty-first century, commonly at the level of the region or metropolitan area as part of an informational, advisory, or visioning effort. While these efforts represent a step toward the accessibility shift, they leave a particular gap at the level of applied decision making about transportation investment and land-use regulation. At the operative level, as when a municipality evaluates a development proposal, or a transportation agency evaluates transportation-investment plans, decisions tend to be guided by mobility-based level-of-service or traffic-impact analysis. The accessibility shift depends on accessibility-based evaluation displacing its mobility counterparts at these operational levels as well as in the regional visioning processes. Yet project-level analysis differs in fundamental ways from regional-scenario analysis of accessibility, and it demands adaptation of the regionally scaled tools (see chapter 3). With accessibility tools scaled to the level of the individual transportation-investment or land-use-regulatory decision, accessibility analysis can be brought to the applied decision-making level.

Comprehensive versus Partial Accessibility

Metrics of accessibility are sometimes judged by their comprehensiveness, with richer description being treated as inherently superior or more accurate than sparser representations. For example, access to health care is not just a matter of people's ability to reach the doorstep of the clinic, but is shaped by their awareness that they need to see a doctor, the kind of health insurance they have, and their ability to overcome any societal forces impeding their ability to get the care they need.[30] Thinking about accessibility in this comprehensive way may make aggregate place-based measures appear to be woefully inadequate descriptors of people's actual lived accessibility. To capture accessibility properly, perhaps only the comprehensive person-based measures, which take into account people's attributes and destination types, will do.[31] This view misconstrues the nature of accessibility, which is not an objectively existing quantity like length or weight to be measured with either greater or lesser accuracy. Accessibility metrics are more like eyeglass prescriptions: they enable users to see things that would otherwise be missed, but only where there is an appropriate fit between the tool and both the user and the user's purpose. Transportation and land-use planners, for example, need guidance regarding transportation investment and land-use regulation to improve physical access to health-care facilities. Health-care obstacles rooted in knowledge, finance, or culture, while vital, are generally beyond their professional purview. The metrics presented in this book (see chapters 3–7) are all geared toward the professional practice of transportation and land-use planning and are hence aggregate and place-based. Other normative or positive uses of accessibility will lead to different methodological choices.

Impact on Complementary System

Accessibility is a function of the interaction of two systems, transportation and land use, that influence each other. A thorough accessibility analysis must therefore allow for the possibility of effects on the complementary system—transportation on land use, or vice versa—before accessibility impacts are gauged. For example, assessing the accessibility impact of land development without accounting for the development's potential effect on roadway congestion would implicitly presume that roadway level of service would be unaffected by the land-use change; the analysis would thus amount to a proximity evaluation. By the same token, assessing the accessibility of a set of transportation investments without first projecting their land-use impact amounts to assuming no land-use change. Incorporating impacts on the complementary system ensures that accessibility analyses properly address both the proximity and mobility dimensions.

Modal Dimension

The accessibility paradigm in transportation and land-use planning has become closely associated with both multimodalism and an equity-based view of transportation. We view the accessibility shift as a potential tool for furthering these goals, but as conceptually distinct from them. The accessibility shift implies planning-practice reform for all transportation modes, including the automobile. An analysis of auto versus transit accessibility, for example, might lead one to conclude that there is an unacceptable deficit in transit service; however, the accessibility shift constitutes reform in the kinds of outcomes that matter for the various modes, not in a fixed position on the priority to be accorded to each. For example, access to the public transit system is sometimes used as a metric of accessibility, an inadequate formulation on two grounds. First, it removes the automobile (and, for that matter, walking and bikes) from the realm of accessibility-relevant transportation modes. Second, it conflates access to the transit system with access to destinations. The bus may run through one's neighborhood, but if job centers are two hours away, work accessibility remains poor. For all these issues, analysis of transit accessibility poses particular challenges, as the partial coverage and intermittent nature of public transit render it a tricky subject for accessibility analysis (see chapter 5).[32]

Equity Dimension

In this way, the accessibility shift does not imply preference of one mode over another, but is an approach to planning that is relevant to all modes. Neither is accessibility synonymous with transportation equity. Logically, accessibility must be a concept distinct from the equitable distribution of accessibility. One expression of this equity fallacy is seen in a tendency to limit accessibility analysis to the transportation disadvantaged, as if accessibility is the service needed by these populations, in contrast to the more affluent car drivers, who need mobility. While car ownership enables higher accessibility than transit use nearly everywhere, accessibility remains the service sought by transportation users across modes and incomes. Accessibility is a resource that is the desired benefit provided by transportation; that resource can be distributed equitably or inequitably. All populations benefit when their accessibility increases, though some populations start from a position of greater accessibility deficit. Under the accessibility shift, transportation analysis focuses on system performance with respect to populations rather than with respect to pieces of infrastructure. The accessibility shift thereby supports analyses of equity issues, which can help direct resources to equity concerns (see chapter 6).

In this book, the accessibility shift is thus quite circumscribed. Rather than asking the traditional question "How fast can people travel?," accessibility planners ask "How many opportunities can people reach?" The shift does not imply that accessibility maximization is the criterion for a successful metropolitan area; accessibility goals necessarily compete for priority with those of other realms such as education, health, or economic development. The shift merely implies that, to the extent that a region seeks to improve its transportation, outcomes will be evaluated (prospectively and retrospectively) in terms of accessibility rather than mobility. Similarly, the accessibility shift does not automatically resolve conflicts among modes; for example, improvements in transit accessibility do not necessarily outweigh improvements in auto accessibility. Similar conflicts may arise between local and regional accessibility, or between the accessibility of various socioeconomically or geographically defined groups. Some areas may choose to ensure a minimum level of accessibility for all populations; other regions might prioritize the accessibility of highly skilled labor to their key industries.[33] Appropriate problem definitions do not automatically resolve planning problems, but logically incorrect definitions, such as those defining transportation's goals in mobility terms, regularly lead to misdirection of planning and policy efforts.

Ultimately, advocacy of both the accessibility shift and its derived-demand logic is grounded in the notion that ideas matter and have the capacity to transform even formidable interests and institutions. Political interests and power undeniably shape policy, and transportation and land-use planners will inevitably incorporate a "muddling through" as they cope with the constraints imposed by budgets, stakeholders' conflicting interests and values, and the natural and social worlds.[34] But the mobility paradigm which currently underpins that messiness in turn shapes the arguments, analyses, and assumptions of planning practice. By altering the terms of debate, the accessibility shift can aid policy reform even in the face of established interests and constraints on decision making. This is particularly so when the ideas are readily transformed into operational tools for on-the-ground decision making within a professional realm, such as transportation and land-use planning, that is well acquainted with the use of analytical methods.

1

WHAT IS TRANSPORTATION FOR?

The derived-demand concept, which underpins the logic of accessibility in transportation and land-use planning, originated in realms entirely removed from transportation. Understanding the framework in general can shed light on its use in transportation and can help eliminate misunderstandings of the meaning of derived demand in the transportation context. In this chapter, we trace the lineage of the concept, its application to the transportation context, and an important challenge to the derived view of transportation demand.

The original example of derived demand—though not the term itself—was the nineteenth-century mathematician Augustin Cournot's analysis of the demand for zinc and copper as a function of the demand for brass, an alloy of the two metals.[1] The derived-demand term was coined in 1895 by the economist Alfred Marshall, who used it to describe the demand curves for goods that were intermediate to the consumption or production of other goods.[2] As an example, he showed how one could derive the demand for knife handles as the difference between the demand for finished knives and the demand for knife blades (neglecting the cost of assembly). Understanding the demand for knives was a step toward estimating the demand for knife handles. The demand for knives was *direct* in that consumers viewed them as a final product; the demand for knife handles was *derived* because they were an input.

This is derived demand in its original form. It is important to note what the term does not imply. First, there is no implication that knife handles can never be used for other purposes; surely some will use them as handles for other implements, as decorations, or as doorstops. But for most consumers and for most purposes, the utility of knife handles depends on their use in completed knives. Second, there is no suggestion that consumers always choose the lowest-cost handle that can do the job. Some knife consumers will prefer handles that excel in their level of ornament, ergonomics, or design. Third, some consumers may even be so enamored of fancy handles that they will accept smaller or less durable blades in order to afford their preferred handle. None of this contradicts the derived nature of demand for knife handles: ordinarily, they are not desired or useful on their own, but as an element of a complete cutting package. And assessments of a household's cutting capability would surely be misleading if they were based on the availability of knife handles rather than knives.

In transportation, mobility is the knife handle and accessibility the knife. The first application of Marshall's derived-demand concept to transportation may have come four decades later in Michael Bonavia's 1936 book *The Economics of Transport*: "The ultimate demand for transport may thus be derived from that for the necessities of life, and be correspondingly inelastic. To some extent, therefore, transport is analogous to other services for which derived demand is very strong although direct demand is weak or non-existent; for example, justice, police or national defense."[3]

Bonavia's grouping of transportation with justice, police, or national defense appears to conflate derived demand with *public goods*—goods or services, such as clean air or national defense, that are freely available to everyone, and for which the use by one person does not reduce their availability to others.[4] In Bonavia's account, the problem with derived demand is that it does not fully reveal itself in a private market. This is not necessarily so. Derived demand, such as the demand for electricity (or knife handles for that matter), may be private and fully expressed in markets. Conversely, the demand for public goods, such as clean air—which markets underprovide—may be direct, rather than derived.

Nevertheless, Bonavia's identification of the derived nature of transportation demand was prescient. It subordinated transportation to "the necessities of life," perhaps a broader formulation than "reaching destinations," which later became central to the definition of derived transportation demand. It portrayed the direct demand for transportation as "weak or non-existent"; thus even the original importation of the derived-demand concept into transportation allowed for a minority of travel that was not for the purpose of reaching destinations.[5]

The derived-demand concept in transportation was developed further by Robert Mitchell and Chester Rapkin, who were interested in forecasting demand for transportation on the basis of land-use patterns across a metropolitan area.

If people travel in order to reach destinations, they reasoned, then the best way to forecast travel demand would be on the basis of observed land-use patterns, which define the origins and destinations of people's trips.[6] Mitchell and Rapkin, like Marshall or Bonavia, were interested in deriving demand for an intermediate good or service—mobility in this case—from the final, or direct, demand for accessibility.[7]

Though Mitchell and Rapkin presumed that the demand for travel was primarily motivated by the desire to reach destinations, they, like Bonavia, did not think that destinations alone fully explained travel behavior. Rather, they argued that destinations can have a range of roles in trip-making decisions. One attribute of a trip was the "determinateness" of its goal, which Mitchell and Rapkin defined along two dimensions: the level of intention to accomplish a goal, and the specificity of the destination. Values for the former dimension ranged from the traveler intending to accomplish "all of a definite number of known goal events" at one extreme to "purposeless existence or movement in a goal area" at the other. Notwithstanding the judgmental sound of "purposeless existence," Mitchell and Rapkin were interested in descriptions of travel behavior rather than assigning inherent values to trips. Specificity of the destination ranged from "a known, intended destination which consists of one or more specific establishments at specific locations," such as a mall or downtown, to a more loosely defined "indeterminate destination, which consists of an indefinitely bounded or centered area."[8]

Similarly, though Mitchell and Rapkin viewed accessibility as a powerful force shaping locational decisions, they were explicit in their view that it would not inherently outweigh all other forces: "Although attention in this study is focused on the manner in which the necessity for movement among establishments influences the local distribution of land uses, this should not convey the impression that the other factors are of little, or even of lesser importance."[9]

Thus, while the purpose of most travel was to reach destinations, the traveler, in Mitchell and Rapkin's view, was no destination-driven automaton. They understood the derived nature of a demand for travel, like other intermediate goods or services, to be a relative, not absolute concept. For Mitchell and Rapkin, it was useful to view most travel as derived because the perspective enabled them to model travel on the basis of the locations of origins and destinations.

The relative nature of derived demand can be understood on a broader level as well. Ultimately, the consumption of any good or service may be conceived as being in service of a higher purpose. For example, food can be viewed as sought not for its own sake but for the sake of nutrition, health, or satisfaction; the line delineating derived from direct demand may be fuzzy. The consensus view that transportation demand is mostly derived is not an absolute truth but, rather, is based on the view that transportation is most usefully viewed, in most circumstances and for most trips, as one means to an end, rather than an end in itself. This proposition can be seen in the asymmetrical relationship between accessibility and mobility: most people would view reaching their destinations with less movement as preferable for most trips, all else being equal. The opposite would not be true: reaching fewer destinations for a given level of travel would usually be seen as a worse outcome.

The Challenge to the Derived-Demand Hypothesis

The derived-demand hypothesis remained the consensus view of the transportation fields in the decades following Mitchell and Rapkin's work, but by the 1990s, some observers were questioning the ostensible absoluteness of the derived-demand framework in transportation.[10] Research in this tradition seeks to demonstrate that, for many people, time or distance spent in travel represents a positive utility in and of itself, rather than a cost (or disutility) that needs to be borne in order to reach a desired destination. This literature effectively explores the multiple dimensions of transportation and the many ways in which travelers adapt to, enjoy, and even derive meaning from the process of getting from point A to point B. It shows the strategies people employ to multitask and hence reclaim some of the value of time spent in travel. It also demonstrates that, for many trips, the destination may be a product of the desire for a trip, rather than the opposite. These challenges to the derived-demand concept have furthered earlier understandings of travelers and travel as multidimensional and layered. At the same time, their analyses help illustrate common misconceptions about derived demand and accessibility, including notions that for transportation demand to be derived, it must be derived for all possible trips; or that the derived nature of travel demand rests on people minimizing transportation costs without regard to other objectives; or that the derived-demand view requires that people dislike travel.

Within this line of research, the derived-demand concept has sometimes been interpreted in an absolutist fashion. For example, Patricia Mokhtarian and Ilan Salomon argue that travel demand is not as derived as is commonly believed, stating that "the strict view of travel as a derived demand would hold that the destination is always 100% primary."[11] This characterization would be consistent neither with derived-demand theory in general nor with transportation analysis in particular. First, nothing in derived-demand theory would require that the directly demanded good is the only purpose to which the derived-demand good can be put; think of Marshall's knife handles. This view does not comport with Mitchell and Rapkin's nuanced view of travel. And researchers have long explicitly allowed for a minority of travel that is for the pleasure of movement, rather than for the purpose of reaching destinations. In general, asserting ironclad laws regarding human behavior would be a risky proposition, and scholars such as Martin Wachs, Gordon Kumagai, and Randall Crane have carefully avoided such claims:

> The demand for movement is a "derived" demand. Movement is *rarely* considered an end in itself but rather as a cost which is normally born [*sic*] in order to achieve other objectives. Except for pleasure driving we normally travel only in order to take advantage of a variety of activities which happen to be separated in space.[12]
>
> For most purposes, a trip is a derived demand, meaning that people typically travel as a means to an end, not as an end in itself.[13]

Thus, the derived-demand view as articulated by Marshall, Bonavia, and transportation researchers following them has not insisted that 100 percent of travel demand is derived, merely a large majority. Going out for a drive, or purposely taking the long route home, have for decades been acknowledged as exceptions to the derived-demand generalization.

The question remains, however: What would count as evidence of a direct demand for transportation, as if time or distance spent in travel were a benefit, rather than primarily a means to reach a destination? The observation that positive attitudes toward travel help explain people's choices to travel greater distances would not, since such attitudes may simply explain their tolerance for travel to more remote destinations.[14] While some people are surely more inclined to travel, Mokhtarian and Salomon's division of the population into "accessibility inclined" and "mobility inclined" segments is not tenable; few ostensibly mobility-inclined people who find that their chosen destinations have moved yet farther away from their home would see themselves better off as a consequence of the change.[15] Similarly, the observation that not all travel choices minimize travel costs does not provide evidence against derived travel demand. For example, the observation that an attribute of a route, such as scenic beauty, is sufficient to induce travelers to spend more time in travel is evidence that travel costs do not outweigh all other considerations when selecting a route.[16] For that matter, travel costs would not be expected to dominate other considerations when choosing a place to live, a restaurant to frequent, or a transportation mode.

To see the fallacy of insisting that derived demand implies travel-cost minimization, consider what such an absolutist view of travel behavior would imply: travelers determined to take the shortest or lowest-cost route regardless of its shortcomings. Nothing, including lack of physical safety, deters them from travel-cost minimization. One might say of such people that they are placing an irrationally high priority on travel costs. Imagining that anything short of that level of determination amounts to a positive utility for travel neglects the vast middle: people who would be willing to increase their travel time or costs in order to realize compensating benefits in other areas. Observations of commuting by choice, that is, travel to work by people who have the option to instead stay at home and telecommute, similarly fail to provide evidence for a broadly held positive utility of travel.[17] Included in the group of commuters by choice may well be people who enjoy the drive to work—just as it may include people for whom the face-to-face interaction with colleagues more than justifies the effort. Thus, observations of this kind shed little light on the question of whether or not the time spent in travel holds a positive utility for large shares of trips.

Similarly unenlightening is the observation that few people value stay-at-home but well-connected immobility, as some authors have aptly explained, presumably under the assumption that this state of affairs represents the epitome of accessibility.[18] People indeed engage regularly in movement, such as the walk in the neighborhood after dinner, that is without a destination in mind.[19] There is undoubtedly a human need to move and explore, as this literature asserts, but physical accessibility—the ability to reach destinations through some combination of mobility and proximity—raises, rather than lowers, the value of those activities. Fulfilling the need to move and explore in an area in which destinations of interest are absent or hard to reach is in general less rewarding, and hence less attractive, than in an area that offers many destinations to be discovered. The notion that the human need to move counters the derived nature of travel demand or the accessibility paradigm is based on the misconception that accessibility's essence is in the capacity to reduce travel rather than in the ability to reach destinations.

Somewhat more systematic evidence of a broadly perceived positive utility for travel comes from survey findings that the desired length of commute is not the shortest possible; for example, when asked about the most desirable length of their commute, median respondents to one survey replied "16 minutes," which suggests that they desire some time separation between home and work.[20] Perhaps a nonzero desired commute time indicates a positive utility for travel.

This interpretation fails, however, on practical as well as theoretical grounds. From a practical point of view, mean travel time to work in the San Francisco Bay Area, the site of the survey, was about twenty-nine minutes at the time of the study; the implication is that the commuter who is worried about too short a commute would be rare indeed.[21] Thus, these survey data do not contradict the finding that, for a large majority of trips and travelers, a reduction in the time they spend in commuting would be a benefit, not a cost. Generally speaking, too-long, rather than too-short commutes remain the problem for individuals and for transportation and land-use planning. This is especially the case in metropolitan regions in which highly productive economies combine with exclusionary municipal land-use policies to constrain commute-reducing options for many people by reducing the affordability of close-in locations.[22]

There is a more fundamental reason to doubt the inference of a positive utility to travel time, even though people seem to desire some degree of home-work separation. Any evidence of such a positive utility would need to systematically net out the value that travelers find at the origin and destination of their trips, lest benefits found at either end be conflated with positive utility for travel. Asking respondents about their desired commute length fails at this task because a zero-minute commute would imply living in the environment of one's workplace, rather than one's own neighborhood. The job might be better accomplished with a question such as this: "Imagine that you live sixteen minutes from work, your desired commute time. A shortcut just opened up which enables you to cut your commute by five minutes. You can take the shortcut and have an eleven-minute commute, or continue how you had been traveling before, for a sixteen-minute commute. There's no difference in attractiveness, interest, or safety between the new route and the old one. What is your preferred route to work after this change?"

A broad preference for the longer commute even in the presence of the new shortcut would constitute evidence against the derived-demand hypothesis. Such a preference might be expected among a minority of commuters who enjoy ancillary benefits, such as some active travelers who viewed their trip as exercise or relaxation as well as transportation. It would be surprising, however, if this were the majority view among drivers and public-transport users. The decisions of where and how to travel and where to live, work, shop, and recreate are a tightly connected bundle; only systematic controlling of the value found at the origin and destination, and along the routes, would be capable of isolating a positive value for the travel itself.

Models of mode choice, route choice, residential-location choice, and destination choice—which are quasi-experimental designs structured to control for other attributes—all confirm that, for most people and trips, travel time represents a cost rather than a benefit.[23] This finding is observed through the negative utilities that are consistently estimated for travel time, and, in the mode-choice case, for the trip as a whole (including its time, cost, transport-mode characteristics, and other attributes) when the value of the destination is excluded. Travelers' motivation for incurring this cost is clear: the benefit they receive from being able to interact with their destinations.

The view that travel time is treated as a cost by most travelers and for most trips does not require people to dislike the process of travel, and research into the positive benefits of travel has revealed the many ways in which people in fact enjoy their travel time. The burden of travel time may at times "translate into becoming a gift for the individual traveler," in Juliet Jain and Glen Lyons's phrase, when it becomes a respite, an opportunity to pursue desired activities, an adventure, or a time for mental transition.[24] The question is the meaning of these benefits for travel behavior and for transportation policy. One possibility is that travelers experiencing such enjoyment would seek more travel over less even if the qualities of their origins, destinations, modes, and routes are held constant. This possibility is contradicted by the families of studies we have mentioned, which reveal that, all else being equal, people seek to reduce, not increase, travel time for most trips. People may enjoy the time they spend in their cars, but taking the long way home merely to have more commute time is the rare exception.

People's capacity at times to enjoy travel appears to conflict with the rarity with which they seek more of it in the absence of compensating benefits in their choice of destinations (e.g., better retail choice), origins (e.g., a more-affordable house), routes (e.g., scenic quality), or modes (e.g., physical exercise). The apparent contradiction suggests a conditional or contextual nature of the benefit that people receive from time spent in travel. For example, busy working parents may enjoy the quiet transition time between work and home, but the legitimacy of—and hence potential for—that enjoyment rests on having to take the commute to begin with. Extending the commute merely in order to benefit from more transition time does not ordinarily work because the condition, having to spend the time, would not be met.

Ultimately, research into the benefits of time spent in travel uncovers much about the transportation (and perhaps human) experience but leaves intact the derived nature of the majority of travel demand that Bonavia found in 1936. People who have the choice will often prefer the remote restaurant that has tastier food or the remote neighborhood that is safer; these observations do not in any way contradict the accessibility-based idea that the more acceptable neighborhoods or restaurants that can be reached—through any combination of proximity and mobility—the better off people are.

From Individual Behavior to Evaluation of Policy

Yet households in the United States and many countries that are able to choose frequently appear to attach little value to proximity when they opt for large homes in remote suburban or exurban locations. One might ask why policy should seek to achieve for people what they do not appear to want on their own. To illustrate the relevance of the accessibility shift even in the face of residential choices that in practice appear to increase, rather than minimize the need for travel, consider first a situation in which residential choices are fixed; that is, people do not have the ability to relocate in response to transportation-system changes. Some householders live in or near their travel-minimizing location; others choose spots requiring travel above the minimum, presumably because location-related benefits outweigh their transportation costs. Consider a transportation investment that would speed up their travel by 5 percent but cause their work and nonwork destinations to shift to places 10 percent more distant. This shift, while improving their mobility, would degrade their accessibility and would leave them with less time and money with which to interact with their destinations. This principle would hold equally for people who had selected neighborhoods that minimized the travel required to reach their destinations and for people who had located farther away.

If the assumption of fixed residential locations is relaxed, the situation changes, affecting different people in different ways. Public investment to speed up travel—even if it reduces accessibility—would leave some people freer to choose larger houses and lots at residential locations farther from their work and nonwork destinations. People with preferences for, and the ability to afford, this kind of consumption—especially those with an appetite for long-distance driving—would benefit from this opportunity; others might be worse off because their destinations would have become harder to reach.

Such a policy would amount to a subsidy for private land consumption, rather than a transportation-improvement strategy.[25] With greater mobility—even with reduced regional accessibility—some people are freer to choose from a greater range of potential residential locations, are able to consume more land for their residence, and in the process, are more able to travel when, where, and how they want.[26] But these gains mask societal losses in terms of degraded ability to reach destinations overall. Moreover, a subsidy to land consumption at the urban fringe would rarely succeed as a political justification for the expense of such investment. Instead, the legitimacy of the infrastructure policy rests on the transportation-system improvement claim that is brought to support it. Under the accessibility shift, the transportation-improvement rationale for subsidies that fail to deliver accessibility gains disappears, and their justification in the face of the significant costs they impose must be sought elsewhere.

The Varying Agendas of the Challenge to Derived Demand

The derived-demand framework in transportation has been criticized from a contradictory set of motivations, one seeming to demote the role of proximity, the other seeming to demote the role of mobility in transportation and land-use planning. Critics questioning the extent to which proximity (or connectivity) can substitute for mobility are motivated by a desire to understand why some people seem to resist so strongly land-use policy aimed at reducing their travel.[27] If significant sectors of the population derive enjoyment from travel, or have an implicit minimum daily travel-time budget which they are seeking to fill, they will not be motivated to choose travel-reducing locations even if these are available, these researchers reason.[28] But if large numbers of people derive a positive utility from travel, rather than seeing it as means to an end, there would be little point in traditional transportation-planning policies such as roadway capacity expansion, whose purpose, after all, is to reduce the time and/or distance people need to travel. Some contend that while time spent in travel may not be a cost for significant numbers of travelers, they do not like spending their time in congestion and would value roadway capacity expansions.[29] Moreover, in this version, without widespread individual motivation to reduce travel time, there is little point to policies such as job-housing balancing, which seek to overcome obstacles to reducing commutes, such as exclusionary zoning.[30] In both these cases, the primary implications of revising the travel-as-derived-demand view are to render proximity-based approaches less relevant to transportation planning.

By contrast, David Banister draws precisely the opposite conclusion from the same challenge to the derived nature of travel demand.[31] His motivation lies in the promotion of sustainable urban forms, which he defines largely in proximity terms. Building on the argument that the demand for travel is not as derived as previously thought, he concludes that accelerating traffic is unimportant. After all, if travel demand is not strictly derived, then travel is a benefit in and of itself. Why focus on speeding up and hence reducing time spent in that beneficial activity? Instead, the implications of the revised view on derived demand, for him, are to reduce the need for travel, encourage shifts away from the private car, and reduce trip lengths.[32]

Neither view logically follows from the challenge to the derived-demand framework. If—contrary to the consensus—most travelers viewed travel time or distance as a benefit rather than a cost, there would be little basis for either mobility or proximity in transportation and land-use planning, since both are approaches to overcoming the spatial separation of origins and destinations. If that separation were a pure benefit for significant shares of the population, they would find nothing in need of overcoming. And since the consensus of the field, including researchers cited here, remains that most travel demand is derived, the challenge to derived demand would provide scant basis for a wholesale demotion of either mobility or proximity in any case.[33]

In sum, the hypothesis that demand for travel is less derived than generally thought has three strikes against it. First, it remains unsupported empirically. In contrast to the derived-demand hypothesis, which draws support from decades of modeling of choices of modes, routes, destinations, and residential locations—and has long allowed for a minority of travel that was for movement—the challenge to derived demand has only speculative evidence. Most of the evidence conflates attributes of destinations, origins, and route-side attractions with a positive utility for time spent in travel. Second, even if the demand for transportation is less derived than previously thought, the finding would not be particularly consequential or transformative for transportation planning and policy. Since a majority of travel would still be derived from people's desire to reach destinations, the derived-demand hypothesis would remain an accurate descriptor for most travel. Third, while the less-derived-than-thought claim has been used to criticize both mobility- and proximity-based policies in transportation and land-use planning, it cannot serve either goal with logical consistency.

Fundamental policy reform lies not in overturning the derived-demand cornerstone of transportation planning, but in building on it. Neither proximity nor mobility is an independently supportable outcome measure for transportation because neither measure, evaluated in isolation, inherently increases people's ability to reach their destinations—which is the core purpose of the vast majority of their travel to begin with. Within this accessibility paradigm, increases in either proximity or mobility are valued as transportation gains only to the extent that they improve either accessibility overall, accessibility by a particular mode, or the distribution of accessibility across society.

Accessibility vs. Travel Behavior in Transportation-Policy Evaluation

Accessibility-based evaluation of transportation and land-use planning is distinct not only from that based in mobility, but also from that based in travel-behavior modification—for example, when reductions in vehicle-kilometers traveled (VKT) are the test of desirable transportation policies. The direct observability of mobility and travel behavior, together with the clear harms with which they are associated, congestion and environmental impact, lend each of them a clarity and urgency as candidate lynchpins of transportation policy making. But because neither is grounded in the core purpose of transportation, neither is adequate to the task of capturing—or even approximating—transportation's benefits.

The travel-behavior orientation implies that the test of transportation and land-use planning is its capacity to reduce (or reduce the growth in) car use. A broadly used evaluation approach in this realm seeks to link individuals' travel patterns to the characteristics of the built environment.[34] Scholars working under this approach hypothesize that areas that are developed in a fashion that is compact, mixed-use, and safe and amenable for pedestrians, cyclists, and public-transport users will influence people's travel behavior toward less driving, with attendant benefits in the reduction of environmental harms (and other realms such as health from increased physical activity). If this causal link can be established, these scholars reason, then policy makers will have the scientific basis they need to advance policies promoting compact cities and regions.[35] Planning for compact development would enjoy a legitimacy rooted in the proven mitigation of environmental and other harms.

Within this framework, accessibility's purpose is to improve travel behavior, and consequently, its test is whether that behavioral modification is achieved. This is an example of accessibility being assessed according to its instrumental role. Under the travel-behavior rubric, absent such empirical proof, there would be little reason to adopt accessibility-promoting policies. This instrumental view of accessibility emerges explicitly in some transportation and land-use research. For example, Randal O'Toole writes:

> Some goals, such as accessibility and walkability, are actually inputs, not outputs. Just because planners judge a neighborhood to be walkable doesn't mean that anyone is actually walking. One plan defines accessibility as "the number of opportunities (such as jobs, shopping, etc.) that can be reached from a given location within a given amount of travel time by auto, transit, or non-motorized modes." This is an input, not an output.[36]

Mokhtarian and Salomon similarly endorse the strictly instrumental view of accessibility:

> The rationale behind accessibility-enhancing policies is that VMT [Vehicle Miles Traveled] and consequently energy consumption and emissions can be reduced by the expected shift of motorized trips to non-motorized modes and to public transportation and by a shift in destination to opportunities in greater proximity to residential areas.[37]

In other cases, the instrumental view is implicit, as with research structured to assess the impact of accessibility on driving, public-transport use, or pedestrianism. Much research at the nexus of transportation and environmental quality since 1970 has in fact focused on the potential travel-behavior impacts of various urban forms.[38]

The focus on the reduction of VKT as the test of progressive transportation and land-use planning is easily understood. Automobile use remains a very large contributor to environmental problems from climate change to regional pollution to local concentrations of air toxics. Notwithstanding technological progress, which has reduced the car's environmental impact per kilometer, rapid VKT growth remains a global threat. Efforts in land-use planning and improvements in public transport, cycling, and pedestrian infrastructure can contribute—alongside technological shifts—to the mitigation of environmental harms associated with the automobile. But in framing the accessibility problem in VKT-reduction terms, the planning field encounters two problems.

First, when VKT is the relevant policy test, uncertainty regarding the presence, magnitude, or statistical significance of VKT-reducing impacts is inevitably construed as an argument against the planning intervention.[39] At the aggregate scale of the metropolitan region as a whole, VKT is fairly closely related to metropolitan compactness; this relationship weakens when the focus shifts to disaggregate studies of individuals and their neighborhoods.[40] Weaker effects at the level of an individual neighborhood are hardly surprising, since travel behavior depends on the characteristics of the metropolitan area at least as much as local neighborhood attributes. Large-scale metropolitan-wide transformations hold great potential for reducing VKT—as evidenced by aggregate studies that compare metropolitan regions with one another—yet any far-reaching transformation will be the product of the aggregation of small-scale incremental change over many years.[41] Though transportation and land-use reformers aspire to the more fundamental changes, their desired innovations are often tested at the local, disaggregate level with attendant controversy about their relevance to transportation problems, a controversy that can stymie action.

The second problem is more fundamental, and more harmful for planning. The VKT generated by a land-use or transportation-system change is a rough metric of harms, or negative externalities of transportation, but it is quite irrelevant to the question of transportation-related benefits. The transportation benefits of such a change would stem from improvements in people's ability to reach destinations, that is, accessibility gains. A framework that seeks to reduce harms is not equivalent to one that seeks to provide (net) benefits, and focusing solely on the harm side of the equation can lead to perverse outcomes.

In particular, judging accessibility as a potential tool to reduce VKT, rather than an inherent benefit, privileges travel that was car-intensive to begin with; without the presence or threat of intense auto use, there are few vehicle-kilometers to be reduced. Suppose a planning agency must decide between routing a new public-transport line to either a high-income, high-auto-use area or a low-income, low-auto-use area. Routing the transit line to the low-income area would have few VKT-reducing effects; even if the line serving the poor area were heavily used, travelers would tend to be diverted from other public-transport lines or nonmotorized travel rather than from the automobile. By contrast, an accessibility analysis would probably find the greatest gains in the poorer area because low car ownership would likely translate into low accessibility levels for these residents.

A focus on VKT reduction as the test of progressive transportation planning in absence of accessibility evaluation can undermine land-use decision making as well. For example, consider a planner seeking VKT reductions from a transit-oriented development strategy. If such a development were targeted to low-income populations with low baseline car ownership and use, any VKT reductions would likely be slight. By contrast, transit-oriented development targeted to an affluent, car-driving population would have a much greater chance of reducing car use significantly. Yet the accessibility gains would be greatest under the scenario in which low-income households were afforded the opportunity to live near public transport; their low levels of automobility imply greater accessibility benefits from the ability to walk to high-quality public transport. For these reasons, without knowledge of accessibility impacts, the VKT-reduction criterion, like mobility enhancement, is inadequate as the central test of transportation and land-use planning.

Nevertheless, the relevance of VKT to environmental outcomes implies that the planner must be centrally concerned about the impacts of decisions and investments on car use, and ideas pertaining to the role of VKT reduction as a policy goal, long implicit in academic research, have seen some moves toward implementation. Beginning in 2013, the State of California initiated a process to reform methods of transportation evaluation used by public agencies statewide. Much of the planning process for municipalities and other authorities in California is governed by the California Environmental Quality Act (CEQA), which mandates reporting on environmental impacts of public decisions together with mitigation where feasible. Consistent with decades of transportation-planning tradition, CEQA had treated degradations of roadway level of service (LOS), a measure of automotive congestion, as an environmental impact to be mitigated.[42] This strictly mobility-based problem definition led to a number of anomalous situations—for example, when CEQA requirements were used to delay implementation of a bicycle-improvement plan because of the threat of degraded auto LOS.[43] To reform this situation, the California legislature passed SB743 mandating an alternative to LOS for the evaluation of transportation's environmental impacts within CEQA.[44] According to the law, the alternative metrics could include "vehicle miles traveled, vehicle miles traveled per capita, automobile trip generation rates, or automobile trips generated." In the process, VKT was designated to fundamentally replace LOS as the central indicator of transportation's environmental impacts within the CEQA process, requiring significant changes to the procedures of municipal governments.[45]

As a metric of transportation's environmental impact, VKT is surely superior to the old LOS metrics, and using it would avoid the problem of penalizing measures to promote walking, cycling, public-transport use, or denser development if they threaten to slow the cars. But the overarching framework of CEQA—or any harm-avoidance approach—is inadequate as a guide to planning because it overlooks the benefits of transportation. A comprehensive system would need to incorporate benefits in the form of accessibility increases while weighing them against harms, which are in part captured by VKT.

In many circumstances, accessibility-increasing strategies would overlap with VKT-reducing approaches. This is because the accessibility shift entails the integration of proximity into transportation and land-use planning. This integration offers the hope of VKT reductions based on the potential for shorter-distance travel and modal shifts—a hope that is backed up by empirical evidence in many cases.[46]

Yet accessibility-seeking and VKT-avoiding approaches are not synonymous. Where the two are congruous, no trade-offs are necessary. This is the case in cell A of table 1.1, which includes cases where an increased proximity from, for example, urbanist development, job-housing balancing, or relaxation of low-density regulations leads to greater origin-destination proximity, and travelers take advantage of that proximity to reduce their travel. By contrast, low-density, auto-oriented development (cell D) tends simultaneously to boost VKT and degrade accessibility, even by car (chapter 4). In both cases A and D, accessibility and VKT goals are aligned. The noncongruent cases include the issue of roadway congestion (cell B), which lowers accessibility by reducing travel speeds, but also constrains trip making and hence VKT. The circumstances that can produce both greater accessibility and greater VKT (cell C) include policies that seek to increase the accessibility of low-income households by providing access to affordable cars, or roadway infrastructure improvements that do not spur extensive low-density, decentralized development. These policies increase automobile accessibility by improving mobility while avoiding triggering a degradation of proximity that would erode or overwhelm these mobility gains.

Perhaps the most perplexing of the cases of increased accessibility together with increased VKT is when proximity gains translate into VKT increases because the reduced cost of travel to the destinations induces more such travel.[47]

TABLE 1.1 Examples of transportation and land-use outcomes with varying accessibility and VKT impacts

	INCREASE ACCESSIBILITY	DECREASE ACCESSIBILITY
Decrease VKT	A. Reductions in car use based on greater proximity of destinations or availability of alternatives	B. Congestion that suppresses trip making Congestion pricing[a]
Increase VKT	C. Proximity increases that induce car trips Cars to low-income households Roadway expansion without high degree of decentralization	D. Low-density auto-oriented development

[a] Notwithstanding its traffic-accelerating effects, congestion pricing raises the time-plus-money cost of travel overall in order to reduce roadway volumes. In absence of countervailing land-use impacts (which might arise) it would therefore be accessibility reducing. Timothy D. Hau, *Economic Fundamentals of Road Pricing* (Washington, DC: Infrastructure and Urban Development Department, World Bank, WPS 1070, 1992), 1–27.

This occurs when more nearby destinations are reachable by car, and individuals respond with an increased frequency of driving trips. The case of increased proximity leading to increased travel illustrates well the relationship between accessibility and VKT-reduction goals. As Bert van Wee shows, increases in proximity that do not translate into travel reductions would nonetheless materialize as accessibility gains.[48] Consider, for example, a case where improved proximity (e.g., from increased density or land-use mixing) affords people the opportunity to reduce their travel by choosing more nearby destinations—and yet they do not. They could have maintained their previous participation in out-of-home activities while shrinking the time and money they spent in travel. By doing the opposite, that is, maintaining (or even increasing) their previous levels of travel, they are demonstrating that their ability to interact with a greater range of destinations (e.g., to shop at better supermarkets, or work at more challenging jobs) is more valuable to them than the potential transportation cost savings that the improved proximity offers. In fact, the accessibility benefits—which are invisible to a planning model focused solely on VKT reduction—are likely to exceed benefits that materialize in travel-reduction terms; accessibility is thus a more robust transportation rationale for reformed land-use planning than VKT reduction is.[49]

The potential for conflict between accessibility and VKT-reduction goals (as embodied in the noncongruent cells B and C in table 1.1) presents a challenge for accessibility-based transportation and land-use planning. Given the importance of VKT as an indicator of the environmental harms of transportation—particularly as these relate to climate change—VKT effects must be a central concern of the field. But the harms associated with VKT effects are properly considered in light of accessibility benefits. The nature of the trade-off relationship would differ according to the values and policy priorities of different decision makers. Nothing in the accessibility shift predetermines, for example, the proper weighting of accessibility concerns for low-income households against the threat of increased VKT, or in general, the relative importance to be accorded to environment, equity, and economic development in policy making. Yet the shift does imply that transportation benefits measured as accessibility are central to decision making, and that these are not adequately captured by a sole focus on either mobility improvement or VKT reduction.

Planning and the Derived Nature of Travel Demand

Mobility, proximity, and VKT reduction have all, at various times, been treated as transportation and land-use planning's central goal. Yet none, in isolation, is capable of gauging the benefits offered by transportation, which are principally in people's ability to reach destinations; for that reason, none, used independently, can meaningfully guide planners' efforts to maintain and improve transportation. And judging accessibility by its capacity to further these goals, rather than as an inherent benefit, neglects the consensus view that the demand for transportation is largely derived from the demand to reach destinations.

Viewing movement as a means to an end, the derived-demand view, does not imply that transportation-cost minimization will, or should, eclipse people's other considerations regarding where to live, where to travel, and how to get there, but simply that the transportation which a choice demands would be one of the costs in the bundle of attributes that the choice represents. The derived-demand foundation on which transportation planning has built for the better part of a century can also support transformations to transportation and land-use planning practice, particularly when closely coupled with on-the-ground (and rubber-meets-the-road) decision making.

2

EVOLUTION OF THE ACCESSIBILITY CONCEPT

The idea of accessibility is not new, either to urban and regional planning or to the social sciences in general. The concept dates back at least to Richard Hurd's 1903 analysis of urban growth, Robert Haig's 1926 "ease of contact," and John Stewart's 1948 "demographic energy."[1] In this chapter, we describe major themes in the evolution of accessibility over the twentieth century, and argue for a grounding of the accessibility shift both in some of the older traditions of the concept and in the derived-demand framework described in the previous chapter.

In Stewart's analysis, the first of the three to quantify the accessibility concept, it was a good predictor of outcomes including observed income at the state level. States with more demographic energy or potential for interaction, such as New York, had much higher income than states with less, such as North Dakota. In this original conception, the primary value of the accessibility concept was positive; here was an independent variable which was useful in descriptive social-science models of phenomena that varied across space.

But if accessibility could predict desirable outcomes, perhaps it could also be a useful indicator for planning. Governments could orient their activities in transportation investment or land-use regulation toward increasing the average level of this apparently valuable resource, or perhaps distributing it more equitably. This would be a normative or prescriptive use of the concept; accessibility could be a societal indicator which public planning consciously strives to improve. Stewart recognized the idea's normative potential and concluded his seminal article with a string of rhetorical questions: "If the maximization of demographic energy or mass is set as a proper social goal, then what part must the freedom of the individual play? Can the sum of human relations represented by $(N1\mu1)(N2\mu2)/d$ [Stewart's accessibility metric] be increased if instead of being permitted to occur spontaneously between pairs of persons the relations are required to be channeled through a chain of social command? Can an individual by taking thought add a unit to his demographic stature? Can society do this for him?" Stewart's answer to these philosophical questions was a final rhetorical flourish appealing to higher authority: "Were the prophets right in insisting that every increase comes as a gift of spiritual grace from a Power above mankind?"[2]

In other words, Stewart recognized the potential of accessibility as a normative goal early on: if energy or accessibility can predict important outcomes such as income, then surely it could also be seen as a policy variable to be directly manipulated by central planners. The reference to the divine might not be publishable in contemporary scientific journals, but it had the virtue of explicitness regarding its normative position: yes, promotion or redistribution of accessibility could be seen as a normative goal, but that would be an unacceptable tampering with the natural order. Accessibility was too valuable to be conferred according to conscious thought or analysis; it had to be accepted as emerging from a source beyond rational human influence. As guidance for policy, accessibility threatened too much governmental coordination. It smacked of a policy agenda extending beyond accepted small-government tasks of establishment of property rights and mitigation of harms to a search for societal benefits. It was simply too hot to handle.

After Stewart's meditation on—and warning against—the potential of accessibility as a prescriptive goal for public policy, accessibility's normative side remained dormant for decades, even as the concept flourished as a positive variable used for prediction and description. Echoing earlier authors, Walter Hansen applied the term "accessibility" to Stewart's "demographic energy" and broadly introduced the concept into the urban and regional realm with three ideas central to the planning use of the tool.[3] First, like Hurd's and Haig's analyses early in the century—and unlike Stewart's nationally scaled research—Hansen's analysis was metropolitan, not the continental. Second, the outcome variable for Hansen was residential development, a central concern of the urban-planning profession. And finally, where Stewart had implicitly treated peoples' inclination to travel as a constant value, Hansen showed that it was a variable subject to empirical investigation. This empiricism allowed Hansen to discover, for instance, that once destination locations and travel speeds were controlled for, people exhibited a willingness to travel greater distances for work than for nonwork purposes.

Positive vs. Normative Accessibility

The strictly positive and predictive nature of Hansen's accessibility analysis was clear from the description of his purposes and thus remained consistent with Stewart's ideas regarding the proper use of the concept. He argued that areas of greater accessibility have greater growth potential, a principle that could be harnessed to develop predictive models of land use.[4] Throughout the 1960s, a growing number of studies built on the accessibility concept for analytical and predictive purposes. These included choice models of residential location (sometimes combined with transportation-mode choice), predictions of traffic flows, and refinements of the accessibility concept itself and its measurement.[5] With data improvements and the advent of advanced tools for spatial analysis, the analytical and predictive uses of the accessibility concept accelerated rapidly in the 1990s.[6]

In the meantime, another use of the concept had become explicit: accessibility promotion as a normative goal, social indicator, or potential metric for the evaluation of plans. In fact, this idea had been present all along. For example, Haig had offered a futuristic vision of the prescriptive use of accessibility in 1926:

> When the city planner of the future is asked whether a button factory belongs on Fifth Avenue, it may be possible to reply somewhat as follows: "This activity consists of 70 per cent fabrication, 15 percent storage, 10 per cent price establishment, and 5 per cent miscellaneous functions, giving it a rating of 2.7 points. Land on Fifth Avenue of this degree of accessibility may not be utilized economically by any activity with a rating of less than 9.3. . . . The button factory should locate, say, in Rahway, New Jersey, or in Stamford, Connecticut." The formula of the future . . . will show what activities can and cannot pay downtown rents. It will supply a scientific basis for zoning.[7]

While Haig offered the scenario in all earnestness, he understood that it was not within scientific reach and ultimately fell back on an externality, rather than accessibility, rationale for land-use regulation: "Unless zoning is fully and skillfully applied, it is entirely possible for an individual to make for himself a dollar of profit, but at the same time to cause a loss of many dollars to his neighbors and to the community as a whole, so that the social result is a net loss."[8]

A more feasible proposal for a normative application of accessibility, five decades later, considered the problem that the evaluation of plans typically occurred through aggregate measures, such as total system travel costs or benefit-cost ratios, which obscured plans' distributional impacts.[9] The proposed antidote to this global level of analysis was a sector-by-sector evaluation of accessibility;

for example, one might model the impact of a plan on the accessibility of low-income workers to employment appropriate to them.

The first reference to accessibility as a normative indicator of transportation-system performance in explicit juxtaposition with mobility is found in an article by Wachs and Kumagai from 1973. Movement, according to them, was not a meaningful measure of policy goals in isolation; more movement could lead to less time and money for other goods or services. Here, they explain the deficiency of mobility-based measures:

> Because the component of expenditure or consumption which is allocated to travel or transportation is not available to purchase the service itself, large travel expenses might result in poorer services rendered or fewer goods consumed. If we study only total expenditure data for particular services we might attribute higher service levels to consumers who spend more money on the service. In fact, large expenditures might result in higher service levels for some, and might be dissipated on transport costs for others. The use of such input information as social indicators may thus be quite misleading.[10]

In distinguishing the resources devoted to transportation from those devoted to a directly demanded service, Wachs and Kumagai articulated a derived-demand rationale for accessibility-based transportation planning. This perspective was echoed several years later by Jennifer Morris, Peter Dumble, and Marcus Wigan with a rejection of mobility ("connectivity" in that paper) measures in favor of their accessibility counterparts for guidance of transportation decisions:

> There may yet be a place for measures of connectivity of the transport network or measures of accessibility to public transport—such measures may be useful in pinpointing glaring deficiencies in the transport system. But for most of the broader issues tackled in present-day transport planning these measures must be rejected on behavioural grounds. Indicators of travel time, distance or cost fail unless supplemented because they reflect only one of the components of the satisfaction an individual may derive from his travel. Account should also be taken of the probable interest of the destination reached.[11]

That study rejected both general mobility metrics and measures of accessibility to public transport (as opposed to destinations) as broad system-wide indicators because they treated movement as the relevant service, rather than access to destinations. Thus, by the 1970s, researchers had identified the inconsistency of mobility-based planning: it was a mismatch with the fundamental purpose of transportation.

The normative view of accessibility was aligned with an evolving understanding in transportation circles that land use was not inherently an external force to which transportation policy responded but the potential object of policy attention in its own right. This evolution was seen in the observation that accessibility planning can improve provision of goods and services through organization of activities in time or location, not just increases in transportation capacity.

Thus, the idea that Stewart rejected in 1948 as a violation of the heavenly order—policy reform directed at enhancing accessibility—was advocated explicitly twenty-five years later, at least on the pages of research journals. Yet, with only a few exceptions, mobility metrics including level of service, travel speeds, and value of time lost in congestion remained the core indicators for transportation policy in the United States and many countries around the world.[12] And the primary professional and research application of the accessibility concept was in its positive or predictive incarnation.

Accessibility Defined as Urbanism

Early work viewing accessibility as a normative goal for planning hewed closely to Hansen's definition of the concept, using indicators such as the number or density of destinations by type that various population groups could reach for a given investment of time or money.[13] The approaches to accessibility that emerged from the positive-measurement tradition inherently incorporated the movement of all modes—even long-distance auto travel—as being captured by the accessibility concept. When articulated in this fashion, there was no need to specify—in fact, no logic in specifying—a separate mobility goal, since accessibility incorporated mobility by definition.

But this particular view of accessibility seemed to wane after initial explorations in the 1970s. It may have been supplanted by the ascendant school within transportation and land-use studies: empirical testing of the link between land-use patterns and travel behavior. By the time the normative view of accessibility reemerged in the 1990s, it had largely acquired a new meaning. The earlier normative perspective had sought to apply metrics incorporating proximity and mobility, the classical definition of accessibility, to the assessment of planning outcomes without presupposing which urban forms would promote accessibility. By contrast, the newer view equated accessibility promotion inherently with urbanist policy prescriptions, such as mixed-use development with a pedestrian, cycling, or transit orientation. The broad concept of accessibility as established by Hansen waned in favor of the urbanist-based focus on local or neighborhood accessibility, or even proximity. For example, Jeffrey Tumlin sharply demotes mobility in the accessibility tool kit, leaving only proximity and connectivity as fully legitimate accessibility tools, and ending up with a focus on local destinations to the exclusion of regional ones: "Rather than focusing on movement, accessibility may bring the product closer to the consumer. Locating a school and a retail main street in the middle of a neighborhood improves accessibility, reducing the need for people to move long distances and improving their choices. Accessibility investments include mixed-use zoning, delivery services, and high-speed internet services that reduce the need for movement."[14]

There are three problems with a focus on neighborhood-scaled proximity to the exclusion of movement along networks. First, congestion of pedestrian, cycling, and public-transport facilities is common in very large cities internationally. This congestion would impede accessibility to local destinations by these modes, but it is not typically accounted for by proximity measures alone. Second, not all local trips are by foot, bike, or transit; auto movements are also relevant to local accessibility, yet proximity without speeds would not incorporate the potential for denser environments to slow auto travel. Third, even the best mixed-use neighborhoods capture only a fraction of nonwork travel.[15] This means that their local accessibility is only a partial indicator of their overall accessibility.

Local accessibility features prominently in many urbanist accessibility definitions. Within this framework, while "automobility" planning strategies have included road construction, intelligent transportation systems, transportation system management, and large-scale public and private transport, "accessibility" approaches include land-use management, telecommunication advances, transportation-demand management, and community-scale public and nonmotorized transport. In general, Robert Cervero argues that "automobility planning works on the *supply side*, aiming to increase the speed and ease of moving about the spread-out city. Accessibility planning, on the other hand, emphasizes *demand management*. . . . It seeks to manage physical space and resources to avoid or minimize motorized travel."[16]

Susan Handy writes about accessibility solutions similarly, denoting options such as road building and intelligent highway system applications as mobility-oriented strategies.[17] These stand in contrast to the accessibility-based approaches including New Urbanism, transit-oriented development, infill development, Main Street programs, and street connectivity.

Defining accessibility in urbanist terms entangles the concept in logical conundrums. This problem can be seen, for example, when large-scale public transport is counted on the mobility side of the ledger, and community-scale public and nonmotorized transport are equated to accessibility.[18] The division is arbitrary, because all urban transportation modes, from walking to cycling to expressway driving, allow the traveler to move and are hence means of mobility. Over the long run, some mobility improvements would be accessibility enhancing; others—because of their capacity to induce a spread of origins and destinations—would be accessibility degrading. But this is a question for empirical observation and modeling, not definition. Identification of a set of built-environment policies that equate to accessibility blurs the fundamental understanding implicit in the classical definition of the concept: rather than a commitment to a particular urban form, accessibility is the ability to reach destinations.

Moreover, if one relies on the classical definition, the very idea of a distinction between mobility and accessibility strategies breaks down. Because mobility is one means to the end of accessibility, the "mobility-versus-accessibility" dichotomy in categorizing built-environment policies and investments makes no more sense than a "proximity-versus-accessibility" trade-off would. To illustrate, consider investments in sidewalks, bike lanes, or urban public-transport lines, which, notwithstanding their identification with an urbanist policy agenda, are surely mobility improvements. In all likelihood, they would enhance accessibility as well, as long as they do not induce extensive spreading of origins and destinations.

The classical definition of accessibility allows greater proximity to be relevant to accessibility even if it does not bring destinations within walking, cycling, or public-transport range. For example, an employment concentration that can be reached only by car but is twenty kilometers away is more accessible than a similar center at forty kilometers' distance. While this kind of improved accessibility might not register on the urbanist's radar screen, it expands accessibility's geographic relevance from zones of walkable urbanism to every corner of the metropolitan region. And by extending accessibility's scope to the automobile, these definitions render the accessibility shift relevant to planning for all transportation modes.

The definitional question is much more than semantic. Beginning in the 1990s, the accessibility paradigm became the language of many advocates of an urbanist planning agenda. While there is surely significant overlap between accessibility and urbanism, failure to keep these concepts distinct ultimately empties the accessibility shift of its analytical content. While many of the urbanists' prescriptions will be accessibility enhancing, a potentially much larger share of accessibility-enhancing policies will not be particularly relevant to urbanism. Moreover, some policies oriented toward smoothing the flow of vehicles, which may include transit, taxis, and other forms of shared mobility, may increase accessibility. As long as accessibility is seen as equivalent to a particular set of built-environment prescriptions, it will have little power to transform transportation and land-use planning. Instead, it will be vulnerable to the charge that it is the domain of a limited market comprised of people with urbanist preferences (and pocketbooks to match), rather than the fundamental purpose of transportation.

Defining accessibility as urbanism seems to lead naturally to a pairing of accessibility and mobility as the twin goals of transportation policy. Even advocates of the shift to accessibility planning pair the two goals in defining the proper operations of transportation planning. For example, Cervero, after urging a shift to accessibility planning, writes: "Of course, all of this is not to suggest that we abandon mobility-based planning. Cars are here to stay, and there remains a need to ensure their safe and efficient passage, albeit in a sustainable and environmentally benign way and in deference to neighborhoods and places. What is called for is more even-handedness. . . . Infusing long-range transportation planning with an accessibility-based mind set provides this."[19] Tumlin, who often equates accessibility and proximity, states similarly that "all transportation systems should invest in both mobility and accessibility, balancing the two."[20]

The pairing of the two goals often occurs with little evidence of the distinction between the two.[21] For example, the 2030 regional plan for the Detroit metropolitan area names as "Goal 1" the objective to "enhance accessibility and mobility for all people" without explaining the distinction between the concepts.[22] CalTrans takes a stab at distinguishing the two but leaves readers scratching their heads as to the distinction: "Mobility refers to the ease or difficulty of traveling from an origin to a destination."[23] "Accessibility is defined as the opportunity and ease of reaching desired destinations."[24] Does the distinction lie in the difference between "traveling to" and "reaching" destinations? Or perhaps it lies in the presence of the word "origin" in the definition of mobility?

Most commonly, however, accessibility and mobility are paired with no attempt at defining or distinguishing them.[25] In U.S. transportation planning, much of this derives from a vague federal mandate: the United States Code specifies that "the metropolitan planning process for a metropolitan planning area under this section shall provide for consideration of projects and strategies that will . . . increase the accessibility and mobility of people and for freight."[26] The relevant sections offer no assistance to the transportation planner seeking to understand and distinguish the concepts, and many metropolitan planning organizations adopt the federally specified accessibility-mobility pairing without meaningful interpretation of the concepts.

The flawed parallelism of the accessibility-and-mobility goal formulation is an unintended by-product of an urbanist definition of accessibility. With accessibility defined in terms of walkability, density, mixed-use development, and public transport, some automobile-relevant transportation-planning goal must be specified. The mobility goal in the accessibility-plus-mobility pairing seems to be a placeholder for the sentiment that walkability is good, but people still need cars, and the cars still have to be able to move. This formulation calling for a balance between mobility and accessibility represents some progress in transportation planning by acknowledging the inadequacy of a sole focus on mobility, but it remains a vague and inconsistent formulation. By treating both terms as goals, this specification fails to acknowledge the asymmetry of the ends-means relationship between them. Moreover, by pairing ends (accessibility) with one means (mobility), the formulation neglects the alternative and sometimes competing means of connectivity and (most notably) proximity.

The consequence of the ungainly mobility-and-accessibility pairing is to limit the transformative power of the accessibility concept. Once mobility enters transportation planning as a parallel goal, metrics of travel speed are seen as meaningfully analyzed in isolation from people's destinations and from the time-and-money cost required to reach them. The superfluousness of the mobility-and-accessibility pairing can be seen as follows: a mobility policy that furthers accessibility would already be covered by the accessibility goal specification. And a mobility policy that degrades accessibility would lower the service people receive from the transportation system. In either case, there is no consistent rationale for an independent mobility goal.

Notwithstanding the argument against accessibility as urbanism by definition, the accessibility shift is highly relevant to an urbanist policy agenda. Mobility-based planning constitutes a barrier to compact, mixed-use, walkable, and transit-friendly development in numerous ways. Land-use regulation that is geared to maintaining highway level-of-service standards tends to limit development in already built-up areas because of the threat of congestion. Transportation investment geared to vehicular movement frequently has the by-product of spurring low-density, auto-oriented areas and limiting walkable urbanist development. In general, proximity, surely one of the keys to successful urbanism, is invisible or worse to mobility-oriented transportation methods. The methods do not constitute an absolute bar to urbanist development, but rather constitute barriers to be overcome. A shift to accessibility-based planning can facilitate urbanist policy options by reducing those barriers.

Evaluative Framework vs. Market Basket of Policies

Though it makes little sense to refer to mobility-versus-accessibility approaches when discussing specific plans, policies, or investments, one can readily evaluate plans or outcomes through alternative lenses of mobility or accessibility. Central to the accessibility shift is not a checklist of urban-development policies or investments but a logically consistent definition of what counts as transportation success and the alignment of analytical techniques and metrics with this definition. When "accessibility" is used as shorthand for "urbanism," its power is undermined. But when it is viewed as the proper framework for land-use and transportation evaluation, its effects could be far-reaching.

This is because evaluation frameworks in transportation and land use are used not only for scorekeeping, but are central to proactive policy guidance. For example, under the common mobility-based practice of traffic-impact analysis, the municipality sets a highway level-of-service standard for its roadways, which serves as a general higher-end threshold for acceptable congestion. The municipality then seeks to regulate new development to attempt to ensure that traffic is not degraded below this standard. A forecast of development-generated traffic worse than the standard would generally trigger requirements for traffic signal improvements, or intersection or roadway widenings. Where these are insufficient to handle the additional traffic load, the municipality may require density reductions in the proposed development. As development and traffic grow, the extent of mitigation needed to maintain automotive traffic flow at the established standard would increase rapidly, and costs could render the development project infeasible. In this way, the practice of traffic-impact analysis, operating as a future-oriented mobility-based standard, can become a binding constraint on the extent of development that can occur in an area. Where this happens, proximity is degraded in attempts to safeguard mobility by automobile.

At times, localities will drop their level-of-service standards in order to allow for denser development, a procedure that illustrates the obstacle that traditional mobility approaches pose to urbanism. Lowering one's standards is rarely seen as a good thing, and politicians who propose such a move are likely to be the target of their constituents' ire. Accessibility-based planning seeks to alter the game by changing the scorecard. Dense development in a built-up area would likely have positive impacts on proximity and negative effects on mobility; the accessibility impacts arise from the combination of the two. The municipality considering denser development would not need to lower its standards in order to approve the proposal; rather, it would evaluate development offering significant accessibility improvements as benefiting, rather than degrading, transportation.

Other Accessibility-Mobility Pairings

Accessibility is sometimes seen as a broadening of the concept of mobility. According to this thinking, traditionally defined mobility—as gauged by highway level of service, for example—is too narrow. But since within this view mobility is the purpose of transportation, the concept of mobility, rather than serving as a means to accessibility, needs to be expanded to incorporate accessibility. For example, a workshop on transportation indicators sponsored by the National Research Council concluded that "a measure of mobility must include elements of accessibility. . . . An example of this type of measure is to record the percentage of travel opportunities within 20 minutes as a measure of what the transportation system offers."[27]

Incorporating accessibility metrics within the suite of mobility indicators represents a broadening of transportation's goals and a partial recognition that movement is not always desired for its own sake. But ultimately, this redefinition of mobility as incorporating accessibility is equivalent to the twinning of accessibility and mobility. It places accessibility in the mix, but keeps simple mobility (as manifest, for example, in highway level of service) as an independent goal, which is inconsistent with the derived view of transportation demand. By subordinating accessibility to mobility, it fails to acknowledge either the end-means relationship between the two or the presence of two alternative means for promoting accessibility.

A related misconception imagines a one-to-one correspondence between accessibility and mobility, that is, mobility is *the* means to accessibility (rather than one of three). This neglects the roles of both proximity and connectivity in generating accessibility. Physical accessibility is always a function of the interaction of mobility and proximity, and never mobility in isolation. For example, the view that mobility is best understood as the unit price of accessibility is accurate in the sense that mobility is an input to accessibility, and it has a cost.[28] But describing the relationship in this fashion is akin to referring to fertilizer as the unit price of crops; fertilizer is one costly input, but only one among several. There are alternative mixes of inputs that can increase the final good or service, and increasing the intermediate input does not always increase the final product. The accessibility shift is meaningful to the extent that it conceptually separates accessibility from mobility; by contrast, understandings that imply that mobility is the sole means to accessibility, or incorporate accessibility into the mobility concept, retain a tight coupling between the two.

Tampering with the Natural Order?

Some obstacles to the accessibility shift are lessened through a more appropriate and precise articulation of the accessibility concept itself. Others may be more stubborn. The implication referred to at the beginning of this chapter, that conscious planning around accessibility somehow amounts to tampering with a celestial order, hints at an important political impediment to the shift. Using accessibility to guide transportation investments and land-use regulations is, to some, an overly intrusive idea. Since accessibility inherently incorporates issues of land use and injects regional-level considerations into what otherwise might be local decision making, the shift may be seen as impinging on municipal regulatory prerogative, something that is zealously guarded in a number of countries, particularly the United States. And since accessibility planning inherently integrates the disparate realms of transportation and land use, it will appear to some as threatening excessive governmental coordination in a fashion that is antagonistic to individual liberty.

Yet both land-use policy—whether in the form of conventional zoning or more innovative instruments of growth management—and transportation infrastructure investments have the by-product of creating, constraining, or redistributing accessibility. No approach to transportation and land-use planning is free of accessibility implications. Few observers of the very earthly planning process would trace the genesis of day-to-day decisions in land use or transportation to a Power above mankind; both the existing distribution of accessibility and its change over time are shaped by planned decisions by human beings, frequently operating within a governmental structure. An accessibility perspective on transportation and land-use planning merely seeks to make the values tied to these decisions an explicit subject of the public discourse: Do planning actions increase or degrade accessibility? Do they advantage accessibility by some modes versus others? Do they render its distribution more or less equitable? Neglecting these questions in favor of the traditional metrics of mobility may succeed in avoiding the appearance of planning intervention into society. But closer examination reveals this impression's illusory nature. Accessibility is unavoidably shaped by policy because land use and transportation are inherently realms of pervasive governmental action. Acknowledgment of this reality demands abandoning the myth that accessibility outcomes emerge from a source beyond human decision making in favor of a conscious linking of the human-made order of the built environment with the human goal of access to destinations.

3

ACCESSIBILITY IN EVERYDAY PLANNING

Consider the conundrum of accessibility-savvy planners in a municipal context. They understand that improving accessibility is the core purpose of transportation and are persuaded that, when viewed as a planning goal, accessibility helps integrate transportation and land-use systems. They seek to use accessibility as a guide for their decision making, but they lack the tools to do it, since the guidebooks and standards that they were trained to follow are almost completely geared toward mobility. For example, the review of a new residential tower on a vacant lot downtown requires a recommendation on whether to approve, deny, or modify the proposal. The city's conventional way of assessing the transportation impacts of new developments is mobility based, using traffic-impact analysis to evaluate the proposed high rise. Downtown streets are already packed, and adding a new residential tower threatens to exacerbate congestion. Yet the project may be an accessibility winner: allowing more people to choose a downtown location as a place to live might offer more in the way of greater proximity than it detracts via degraded mobility. Conventional mobility tools may steer the decision toward shrinking the proposed development to mitigate predicted congestion, without ever considering its potential accessibility gains.

To make matters worse, currently available accessibility metrics seem ill adapted to the problems these planners face, because the existing tools provide an accessibility snapshot for an entire metropolitan region, rather than a gauge of the incremental effect of an individual project. Yet individual projects, whether transportation or land-use based, form the context of most applied planning decisions. To advocate for an accessibility perspective in decisions like these, planners need tools that shift from mobility-based to accessibility-based analysis and work at the project scale as well.

In this chapter, we develop methods to support accessibility-based analysis for both land-development and transportation projects to help forge a closer link between accessibility analysis and applied decision making in planning. The chapter begins with a brief survey of methods for accessibility measurement and then demonstrates empirical application at the project level using data from real-world land-development projects in Ann Arbor, Michigan, and a suite of transportation projects in San Antonio, Texas.

Measuring and Representing Accessibility

Accessibility metrics vary in their measurement approach, purposes, and levels of complexity.[1] This section summarizes the foundational concepts of accessibility measurement and representation. Accessibility is normally reported in the form of a score or index to describe the ease of reaching destinations from a place, which allows analysts to compare accessibility from one place to another, or track changes in accessibility over time. Sometimes an accessibility score conveys a directly interpretable meaning, as with the statement "1,000 jobs are within 15 minutes of travel by car from Zone A." Sometimes an accessibility score is designed to be a dimensionless index, so that the index only has a relative meaning for comparison between areas. Analysts face several choices when measuring accessibility, but most approaches include a common set of elements.

Origins

The accessibility scores used in this book are attributes of places, so specific place locations must be defined to establish an accessibility measure. An origin is typically defined as a zone, such as a traffic-analysis zone, census tract, neighborhood, or block. But the place can also be a point in space, such as a home or a particular building's address. Analysts often use attributes of the origin, such as the number or characteristics of people living there, to link place-based accessibility to particular types of people.

Destinations

The places that travelers seek to reach also require definition. Because of the importance of employment, jobs are a common destination type in accessibility analysis. Destinations can be disaggregated into categories, as when an analysis divides jobs into various categories based on wage levels, and then calculates separate accessibility scores for each wage level.[2] Nonwork destinations involve a range of activities, such as shopping for groceries, studying at school, or consulting about health care at a medical center. An analysis might combine various destination types into a single, composite metric of nonwork accessibility, as discussed in chapter 7. Other nonwork analyses calculate separate accessibility scores by sector, such as one for grocery stores and another for schools.[3] Weighting destinations by their relative attractiveness is important. Weights can be constructed on dimensions such as building size, product or service affordability, trips generated, or service quality. Like origins, the identification of destinations is typically in the form of a structure of zones but can also be individual points in space.

Travel Cost (or Impedance)

Accessibility scores account for the cost of travel from origin to destination. The most common approach is to use travel time, and that is the approach used in this book, but other costs can also be used, including distance and/or monetary expenditures. Sometimes *generalized cost*, a combination of time and money costs, is used as the indicator of impedance.

Impedance Function

While the cost of moving from origin to destination is directly measurable, its meaning varies across travelers. Some would consider a thirty-minute nonwork trip to be short, while others would see it as quite long and would behave accordingly when choosing how and where to travel. Analysts use an impedance function to capture the increasing difficulty of interaction as travel costs increase.[4] The relationship between travel costs and perceived difficulty of interaction varies by trip purpose. For example, travelers feel a greater "friction," or resistance to longer distances, when setting out to buy groceries than they do when going to work. This phenomenon is captured via differing formulations of the impedance function or values of the parameters it uses.

Mode and Times of Day

Because travel costs depend on the mode of travel (e.g., a bus typically requires substantially more time than a car to travel the same distance) and also on the time of day (e.g., morning and afternoon peak periods require more travel time), accessibility scores are typically calculated separately for each mode, and sometimes for different periods of the day (e.g., peak and off-peak periods). Modes also are influenced differentially by travel distance, which is reflected in differing modal impedance functions; for example, the difference between one and two kilometers is hardly relevant while driving, but the impedance experienced between those two distances would be substantially different for a pedestrian.

Approaches to Measuring Accessibility

Accessibility metrics differ in their handling of these elements, especially in the treatment of impedance. The simplest technique is the cumulative-opportunities approach, a straightforward count of the number of destinations that can be reached within a given travel time or cost. To illustrate, if zone A has 10 restaurants within a 20-minute drive time, and zone B has only 5 restaurants within a 20-minute drive, zone A has twice the restaurant accessibility by car of B.[5] This approach has the advantage of providing an accessibility score that is easy to understand. The main weaknesses are that all destinations within the fixed threshold are weighted equally even though distant destinations are less desirable, and all destinations beyond the selected threshold are ignored.

Gravity measures of accessibility overcome these weaknesses of the cumulative-opportunities measure by weighting close opportunities more heavily than those farther away.[6] But this improvement in measurement comes at the cost of losing the score's intuitive meaning, since gravity models produce an accessibility score in the form of a dimensionless index.

Utility measures of accessibility offer a third approach to handling travel costs and impedance. Utility-based calculations of accessibility are based on discrete-choice models of destination choice and are a gauge of the expected utility provided by a choice of available destinations and modes.[7] Such utility-based measures of access to opportunity are able to account for how different individuals value elements of their journey differentially. For example, some travelers might prefer low-cost but more time-consuming trips while others might value faster trips even if they are more expensive. By capturing such differences, utility metrics excel at estimating accessibility at the level of the individual, an important advance considering the many person-specific characteristics that determine people's ability to interact with their environment. However, utility-based accessibility calculations are considerably more complex to calibrate and interpret, a factor that limits their applicability in applied planning practice.

In this book we employ gravity metrics in most instances, but with no claim to their inherent superiority. In general, accessibility metrics are employed for purposes that vary between contexts; this implies that no metric is superior in all instances. In particular, transportation and land-use planning's future orientation tends to limit the applicability of complex utility-based accessibility metrics. This is in part because the identity, preferences, and travel constraints of future residents are usually unknown. For this reason, the cumulative opportunity or gravity methods may be more applicable to many future-oriented planning decisions.

Representation of Accessibility Metrics

Accessibility metrics may be represented as a map showing accessibility values across zones, a chart displaying accessibility's distribution across a population, or a summary score that aggregates multiple accessibility scores into a single value for an entire city or region. Also of interest may be the change in accessibility across time or space, as when an analysis compares across time periods or alternative development scenarios.

When metropolitan job accessibility is displayed in map form, most metropolitan areas illustrate a centralized accessibility pattern, with accessibility highest in the central business district and declining slowly outward in all directions. This finding holds despite large-scale job decentralization in the United States, or the location of employment in many subcenters and the areas between them. Even in regions where the central business district of the largest city holds only a small share of regional jobs, it generally remains the accessibility high point.

There may be pockets of concentrated access located in regional subcenters, but these subcenters tend to have relatively poor access to the other subcenters, whereas the central business district has reasonably good access to nearly everywhere in the region. The decentralized Los Angeles region exemplifies these phenomena, with the highest employment accessibility by car observed in the central city, and decreasing accessibility bands headed to the suburban regions of Los Angeles County and its neighbors (figure 3.1).

Public-transport accessibility is more geographically irregular than its auto counterpart, because transit accessibility tends to be concentrated around station areas, and so sometimes displays low points in areas between stations.

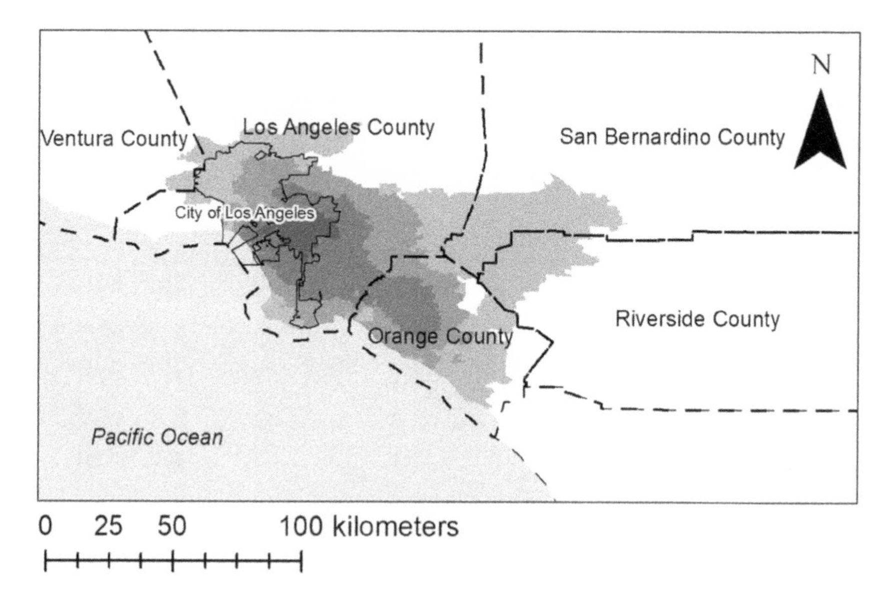

FIGURE 3.1. Distribution of transit accessibility for metropolitan Los Angeles, 2012

The disadvantage of the map format is that is does not tell the story of how accessibility is experienced by residents; for example, a large area of high accessibility may have sparse population and hence affect people's ability to reach destinations only slightly. A chart can overcome this drawback by showing how accessibility is distributed across the entire population, as well as how it is experienced by a particular population segment.[8] For example, transit accessibility varies markedly across the Los Angeles region, with virtually none for the 20 percent of residents experiencing the lowest accessibility (figure 3.2). The 90th-percentile resident enjoys about 30 percent of the transit accessibility of those who live in the center of the region and experience the maximum transit accessibility the region has to offer.

At the level of the region or city, the population's average accessibility can be calculated as a single number for purposes of intermetropolitan comparison or benchmarking performance over time.[9] This is typically accomplished by taking a population-weighted average across a set of zones. This can allow an analyst to conclude, for example, that San Francisco has a higher average public-transport accessibility than Washington, DC, or that accessibility to jobs by bicycle has improved 12 percent from 2005 to 2015. The quantification of accessibility at the aggregate scale hides a great deal of specific demographic and geographic variation but can be useful in comparisons of regionwide accessibility performance.

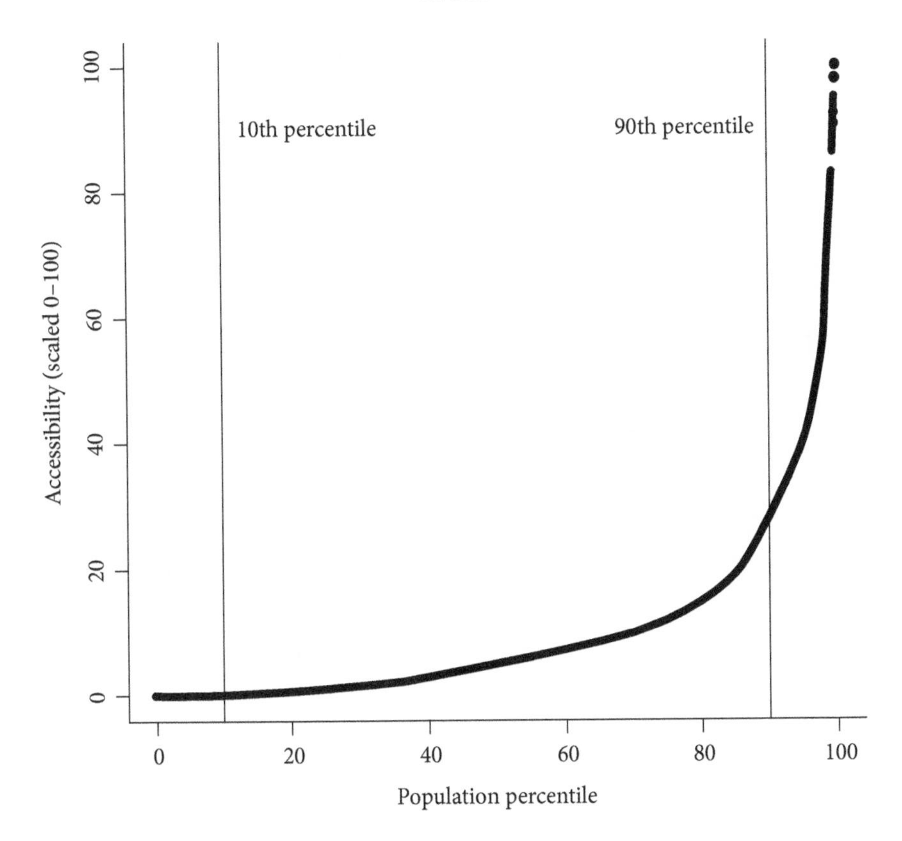

FIGURE 3.2. Transit accessibility to jobs by census tract for Los Angeles metro, 2012

Accessibility Evaluation in Regional Scenario Planning

This regional scale was the first place accessibility made its mark in applied planning practice through scenario-planning efforts, which lend themselves readily to accessibility measurement. Regional scenario planning is a natural venue for the application of accessibility metrics because scenarios consider complete and well-defined potential land-use and transportation futures. In this context, accessibility metrics are a succinct way to evaluate how well the transportation and land-use systems work in combination to facilitate people's ability to reach desired destinations.

Regional scenario plans are based on hypothetical futures for metropolitan areas.[10] Several possible futures are sketched out, including varying land-use patterns and alternative future regimes of transportation infrastructure. These futures scenarios are typically organized thematically, often with one representing "business as usual" and others representing potential counterpoints to current development patterns, such as increased development in centers or along major public transport corridors. The scenarios may project a state twenty or more years into the future. A variety of performance measures are then applied to compare this small set (typically about three to five) of future scenarios. Performance measures may address a range of issues including economic development, affordable-housing supply, environmental quality, and the functioning of the future transportation system. With respect to transportation performance measures, such scenario plans typically consider mobility in terms of the level of congestion in the regional roadway system and/or delays in regional travel times. In a few cases, accessibility-based performance measures have been applied in regional scenario planning as well.[11] However, these accessibility metrics rarely displace their more commonly used mobility-based counterparts.

The accessibility metrics most commonly applied in regional planning are simple and easy to interpret. They are typically cumulative-opportunity-based measures of residential access to jobs, such as the average number of jobs that area residents can reach by auto within thirty minutes. Some metropolitan areas disaggregate their accessibility to jobs measures by population group or by mode.[12] Disaggregating by population group allows analysts to assess how various scenarios perform with regard to equity. Disaggregating by mode also enables analysis of multimodal accessibility, or how future scenarios perform with respect to public-transport, auto, and active-transportation accessibility.

Regional-scenario planning serves an important purpose by offering a big-picture view of transportation objectives and priorities at the metropolitan scale, but frequently lacks explicit links to implementation. Regional change unfolds over many years as the product of thousands of individual land-use and transportation decisions, each of which contributes only marginally to changes in the population's regional accessibility. Regional-scenario planning by itself is inadequate for everyday planning decisions for two main reasons. First, regional-scenario plans are generally advisory and are not inherently linked to implementation, at least in the United States. Instead, such scenario plans promote an overall regional vision, which local governments then implement to varying degrees at their discretion.[13] Second, regional-scenario plans present a limited and purposefully simplified choice set. They offer only a handful of regional alternatives that are all presented at a single future point in time, removing the dynamic complexity of the transportation–land use relationship. Missing from this limited set of scenarios are the impacts of hundreds or thousands of small, incremental changes that are neither prohibited by nor required by any specific scenario.

From Regional Analysis to Project Analysis

Moving accessibility-based evaluation from the regional scenario to the project level involves more than applying regional tools to individual decisions, because project-level analysis is inherently different from a regional analysis. Part of the difference is in geographic scale, since an individual land-use or transportation project is nearly always less than regional in its scale. But the more important difference is a shift in the question being asked. Regional scenarios of transportation/land-use systems, whether current assessments or forecasts of potential futures, are complete and well-defined states at a fixed point in time. Thus, regional-scenario accessibility analysis amounts to a snapshot in time that inherently captures both transportation and land-use aspects.[14]

By contrast, project-level evaluation begins as an analysis of a *marginal change* in either the transportation system or the land-use system, and on rare occasions, in both. The sector-specific nature of the change complicates its evaluation because a change to land use, for example, also has impacts on the transportation system, which must be forecast. Furthermore, whereas future alternative scenarios are readily compared with one another, the basis of comparison for a marginal change in land use is not self-evident. In addition, projects are often quite small in relation to the regions of interest. This scale difference can make accessibility analysis more challenging than when analyzing dramatically differing regional futures.

Attributes of Project-Level Accessibility Evaluation

Project-level accessibility evaluation differs from regional-scenario analysis in three important dimensions. These dimensions vary somewhat depending on whether land-development or transportation-investment projects are being considered.

COMPLEMENTARY SYSTEM

The impacts of projects on the complementary system must be forecast before they can be analyzed. That is, land-development plans require an analysis of their mobility (or traffic) impacts, and transportation plans need analysis of their land-use change impacts. For project-level analysis, neglecting the impact of transportation on land use or vice versa amounts to an implicit assumption of "no impact." The land-use effects of a transportation investment pose a modeling challenge because such changes take time to materialize. Nevertheless, anticipating the land-use impacts of a transportation investment is essential to a meaningful analysis of accessibility because, under the implicit assumption of "no land-use impact," all mobility improvements become accessibility improvements. Only when the possibility of induced development is introduced do accessibility and mobility become truly separate measures. In the evaluation of land-development projects, the forecasting step faces other challenges. Although such projects are regularly evaluated for their mobility impacts via traffic-impact analysis, these analyses are highly localized, so their regional impacts are not readily apparent.

BASIS OF COMPARISON

The second attribute required for project-level evaluation, particularly as pertains to a land-development project, is a meaningful basis of comparison for each project under evaluation. Should it be compared with the same amount of development in a different location? If so, which locations should be considered? Analyzing the accessibility impact of a proposed development is also made more difficult because the identities of its future residents are unknown, as are the locations from which they would move, and the locations they would choose if the project were not developed. For this reason, it can be unclear what alternatives a particular land-development proposal should be compared to in order to gauge its accessibility impacts. Transportation projects, frequently the product of direct governmental investment, differ from land-development projects in this regard. Transportation projects are typically drawn from a long list of potential projects, only some of which can be funded. Therefore, they can more readily be compared to one another with regard to their accessibility impacts. Also, it has long been customary in the transportation field to compare build and no-build scenarios for proposed transportation projects.

SCALING TO REGIONAL IMPACT

A final challenge more particular to land-use projects stems from the size of the project relative to the size of its region. Individual development projects are drops in the ocean of regional accessibility, yet they can add up to substantial change as such projects accumulate over time. Some transportation projects are also quite small and may have barely noticeable effects. However, transportation projects amenable to accessibility analysis are typically much larger in scale and therefore have more readily discernible impacts on the availability of regional destinations. Since an individual land-development project will barely affect region-wide accessibility, analyses must be scaled to the magnitude of the project in question.

Evaluating the Accessibility of Land-Use Developments

A new land-use development creates two impacts on accessibility patterns, one via proximity and the other via speed. A development's residential component can be thought of as new origins, and any commercial component can be thought of as new work or nonwork destinations. Therefore, the development's accessibility impact depends in part on its physical location. For example, if a new residential development is allowed in an area with high accessibility, then it is likely that the development will result in greater average accessibility overall for the region, as long as its impact on traffic is not too severe. Likewise, allowing new destinations in a location that is currently sparse in destinations can also boost regional accessibility. The speed impact of a new development results from its users adding vehicles onto the road network, thereby slowing down traffic, an impact that is traditionally analyzed via traffic-impact analysis. While the method developed here builds on traffic-impact analysis for its mobility component, it is intended to show how analysis can go beyond a focus on mobility to become a true accessibility-based evaluation.

Under this approach, land-use change is derived from the size of the project itself, gauged here in terms of population for residential development and employment for commercial development. This land-use change affects the complementary system (transportation) in the form of additional traffic, analyzed here by importing the results of a traditional traffic-impact analysis. Such analyses forecast additional delays, denominated in seconds, to intersections near the proposed developments. These delays, together with the land-development proposals themselves, are then used to calculate accessibility impacts.

The key innovation of the method is the measurement of accessibility impacts through the *accessibility elasticity*. Accessibility elasticity is defined as the percentage change in regional accessibility divided by the percentage change in regional size associated with the proposal. It is calculated with respect to population for residential developments, and with respect to jobs for nonresidential developments such as retail or commercial.

The accessibility-elasticity metric solves two problems: establishing a basis of comparison and scaling to regional impact. The implicit basis of comparison of the accessibility-elasticity metric is the existing built environment in the region in question. An accessibility elasticity of 1.0 indicates that the marginal person or job at the contemplated location contributes to the total accessibility of the region equivalently to the average current regional resident or job. Elasticities of over 1.0 indicate that incoming people or jobs are contributing more than the average; elasticities of less than 1.0 indicate that the contemplated project degrades average accessibility relative to the existing pattern of development. This is much simpler than, for example, comparing a proposed development with an equivalent development in other locations, which would involve performing complex additional analyses to establish plausible hypothetical alternatives.

The second problem that accessibility elasticity solves is that of scaling to regional-level impacts. A single, small-scale project scarcely moves the regional accessibility needle at all. But since the percentage change in regional population or jobs is used in the denominator of the accessibility-elasticity calculation, the metric is calibrated to compare the effect of even small marginal changes. The accessibility-elasticity metric neutralizes the effect of the development's size, so both small and large developments can be compared on an equivalent basis.[15]

Step-by-Step Procedure

The procedure is illustrated here with data from three proposed developments in Ann Arbor, Michigan, shown in figure 3.3.

ARBOR HILLS

This is a high-end retail development located about six kilometers from downtown Ann Arbor along the major east-west corridor of Washtenaw Avenue. It includes 8,400 square meters of retail space with 692 estimated trips occurring during the afternoon peak hour. Traffic-impact analysis was conducted in May 2011.

413 EAST HURON

This is a residential tower located in downtown Ann Arbor including 216 residential units and 450 commercial square meters on the ground floor (the impact of the commercial is not included in this accessibility analysis). Peak trips for this development were estimated at 112 for the 4:45–5:45 p.m. hour. Traffic-impact analysis was conducted in 2012.

NIXON CONDOMINIUMS

This is a moderate-density residential development of 473 units located about six kilometers from downtown Ann Arbor, contiguous to existing single-family and multifamily residential developments. Trips were projected at 265 at the 8:00–9:00 a.m. morning peak period. Traffic-impact analysis for this project was conducted in 2014.

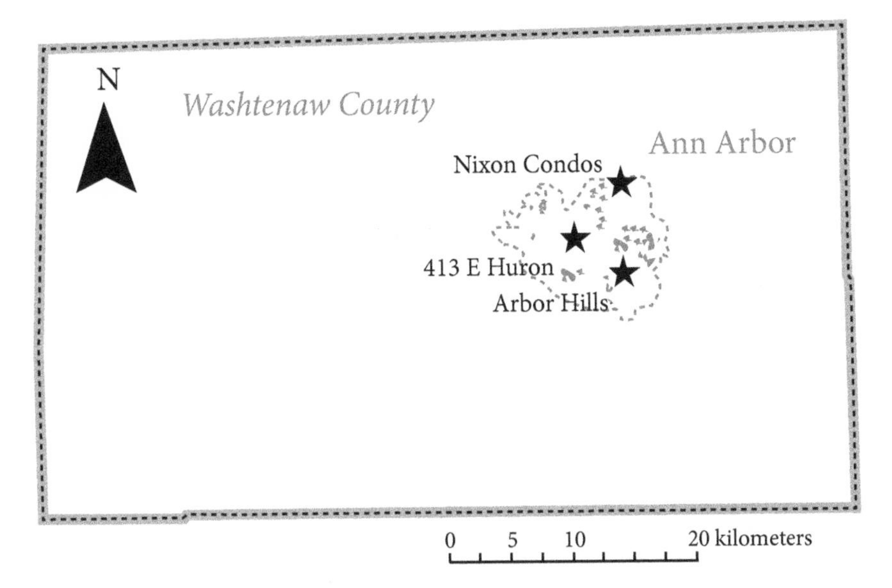

FIGURE 3.3. Locations of land-use development projects in Ann Arbor, Michigan

There are several analytical steps to implementing the accessibility-elasticity metric.

CALCULATE TOTAL REGIONAL ACCESSIBILITY

For this case study, we employed a gravity-based accessibility formula to measure total countywide accessibility to all employment for Washtenaw County, Michigan. The Southeast Michigan Council of Governments provided population and employment counts by traffic-analysis zone (TAZ) and auto travel times between zones by time of day: a.m. peak, p.m. peak, and off peak. Based on research that suggests a relationship between metropolitan population size and the work-travel impedance function,[16] a range of impedance coefficients typical for work travel for smaller metropolitan areas was used. Total regional accessibility is the sum of accessibility from each zone multiplied by the resident population in each zone. Total accessibility, rather than average accessibility, is used in this analysis because the effects of population growth are incorporated at later stages through the elasticity calculation.

CONDUCT A TRAFFIC-IMPACT ANALYSIS

Based on traffic-impact analyses conducted for the projects, an average delay for each surrounding intersection is calculated.

UPDATE REGIONAL TRAVEL TIMES

Intersection-level delays from the previous step are assigned to zone-to-zone travel times via a routing procedure.[17] This is completed by calculating the route from every zone to every other zone and identifying which routes run through the affected intersections. This computationally intensive procedure is implemented with programming and makes use of open-source mapping and routing data. The result is a new set of zone-to-zone travel times which accounts for the delays created by the proposed development.

UPDATE REGIONAL LAND USE

Given the proposed development's location and size, either population or employment is added to the appropriate zone. Housing units are changed into a population change based on average household sizes for each particular type of housing unit within the region. To convert retail development size into retail employment, the Planners Estimating Guide suggests that each retail job requires approximately 47 square meters (510 square feet) of space.[18]

CALCULATE THE ACCESSIBILITY ELASTICITY

With new forecast information on land-use and transportation patterns that will result from a project, a new, postdevelopment total regional accessibility is calculated. The same gravity accessibility formula is used, but with new travel times, new population counts, and new employment (as destination activities) information.

Accessibility elasticity is computed as the proportional change in accessibility divided by the proportional change in regional size (see equation 3.1). The accessibility elasticity score minus 1 yields the percentage by which the proposed development outperforms or underperforms the current average accessibility of the region.

$$accessibility\ elasticity = \frac{\left(access^{after} - access^{before}\right) / \ access^{before}}{\left(population^{after} - population^{before}\right) / \ population^{before}} \qquad (3.1)$$

Accessibility Results for Proposed Ann Arbor Developments

The results of the accessibility evaluation of the three land-use development projects indicate that two projects exceed the average-accessibility threshold for Washtenaw County: 413 East Huron (1.86) and Arbor Hills (1.49), with 413 East Huron providing somewhat higher accessibility overall (table 3.1). The Nixon Condominiums project has an elasticity of less than 1.0, which indicates that an

TABLE 3.1 Accessibility analysis of land-use developments in Ann Arbor to jobs by auto

PROJECT NAME	NIXON CONDOMINIUMS	413 EAST HURON	ARBOR HILLS
A. Development type	Population (residential)	Population (residential)	Retail employment (commercial)
B. Analysis area	Washtenaw County	Washtenaw County	Washtenaw County
C. Impedance coefficient	0.13	0.13	0.28
D. Time of day	AM	AM	PM
E. Accessibility before	7.439E+09	7.439E+09	1.020E+08
F. Accessibility after	7.448E+09	7.452E+09	1.035E+08
G. Percent change in accessibility [(F/E)–1]*100	0.125%	0.182%	1.455%
H. Total baseline before	344,791 (population)	344,791 (population)	19,466 (retail jobs)
I. Total baseline after	345,529 (population)	345,128 (population)	19,644 (retail jobs)
J. Percent change in baseline [(I/H)–1]*100	0.214%	0.098%	0.914%
K. Accessibility elasticity (G/J)	0.59	1.86	1.49

incoming new resident would contribute less accessibility than the regional average. Each resident of 413 East Huron would contribute auto accessibility to work at a rate 86 percent higher than the average resident of Washtenaw County. By contrast, each resident of Nixon Condominiums would contribute accessibility at a rate 41 percent lower than the county average. As a retail facility, Arbor Hills is evaluated on the basis of the jobs it is expected to contain; each job at that site contributes accessibility at a rate 49 percent greater than the average retail job in Washtenaw County.

Depending on circumstances, planners may have good reason to advocate for projects that result in elasticities of less than 1.0. Indeed, based on this small sample, it is clear that the 1.0 threshold is relatively difficult to reach for a new development, except in locations quite close to the urban center. The necessary level of accessibility elasticity for development approval will require analysts to make judgments tailored to local circumstances.

Beyond Traffic Impact Analysis

Projects like the ones shown here all result in added forecast roadway congestion; local jurisdictions typically need to decide, after a traffic-impact analysis, whether the cost of worsened congestion is at an acceptably low level. The proposed accessibility elasticity method goes beyond the mobility-only focus of traffic-impact analysis to show not just the cost of congestion but also the countervailing benefits of enhanced accessibility.

Development approval must consider multiple factors beyond accessibility alone, such as the adequacy of infrastructure, the quality of the urban design, and consistency with current zoning. However, to the extent that transportation is the issue at hand, accessibility impacts should be the primary benchmark by which transportation-related impacts are assessed. Although the case study demonstrated here is focused on automobile-based accessibility, the method can be extended to other modes as well. A thorough accessibility evaluation for a proposed development project would take into account the accessibility impacts for all modes and consider the importance of each mode according to local planning priorities.

Evaluating the Accessibility of Transportation Projects

Like the accessibility analysis of land-development proposals, the analysis of transportation investments' accessibility requires a forecast of impacts on the complementary system (land use in the case of transportation projects). Forecasting impacts of transportation investment on land use is more complex than the reverse, however, because land-use impacts materialize over a longer time frame and also because there is greater uncertainty in the land-use forecasting process. The case study that follows demonstrates accessibility-based analyses for transportation projects and compares the results to a pure mobility-based analysis.

The accessibility-based evaluation ultimately comes to different conclusions than an evaluation based solely on mobility impacts. Transportation projects have two effects on accessibility, one short term and direct, and the other long term and indirect. Traditionally only the short-term, mobility-based impacts have been considered in transportation plan and project evaluation. The short-term impact is the reduction in travel times, costs, or distances resulting from the new infrastructure; this speed component increases accessibility by reducing generalized travel costs. The long-term impact is from the induced land-use effects,[19] which may counteract these short-term mobility-based benefits through land-use dispersal, or, alternatively, enhance them through land-use concentration.

Transportation analysts commonly take the future spatial distribution of land uses as fixed and exogenous to the transportation planning process to allow them to predict trip patterns. This may be due to a number of factors, including lack of technical capacity, since few metropolitan planning organizations maintain integrated transportation/land-use models; the politicized and hence hard-to-predict nature of land-use decision making; and the desire to garner resources for infrastructure investment without calling attention to undesirable land-use impacts. Taking future land-use patterns as exogenous, that is, unaffected by transportation investments, biases forecasts of the accessibility impacts of transportation projects. Because transportation projects do shift future land-use patterns, land-use shifts must be taken into account within any comprehensive accessibility analysis of transportation projects.

The following case study illustrates an approach to accessibility-based transportation project analysis that incorporates the potential for induced land-use change, employing a land-use model to forecast how proposed transportation projects are likely to affect employment and residential locations. This contrasts with the current state of the practice (figure 3.4, bottom) in which a future land-use forecast serves only as an input into shaping future travel patterns.

The proposed method addresses all the requirements for a project-level accessibility analysis. Impacts to the transportation system and the land-use system are forecast through a travel-demand model and a land-use model, respectively. In this case, the land-use component of accessibility is calculated based on forecasts of the locations of jobs by industry and households by income category in future years. The basis of comparison for transportation projects here is the no-build scenario.

The scale of project that is analyzable in accessibility terms differs between the land-development and transportation contexts. Any land-development project that is amenable to traffic-impact analysis can be analyzed for accessibility impacts using the approach described in this chapter. By contrast, a transportation project

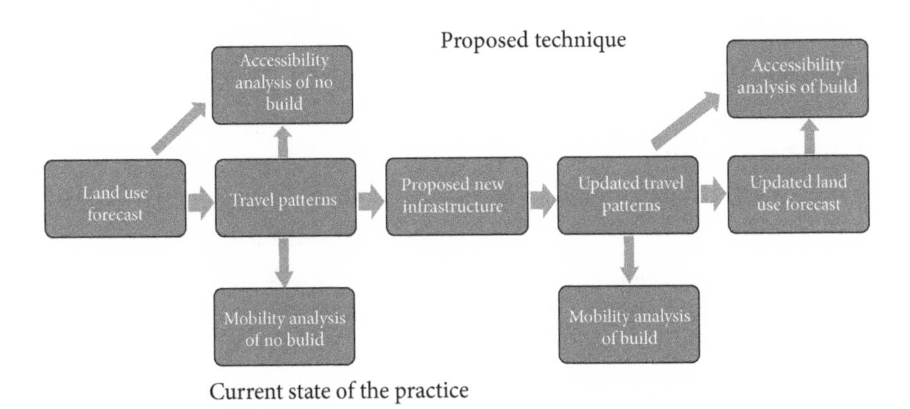

FIGURE 3.4. Accessibility-based transportation project evaluation

would need to be significant at the regional scale before its land-use impacts could be meaningfully forecast. Individual transportation projects that fall below the threshold of discernible land-use impacts could be bundled into larger plans for analytic purposes. We have used this latter approach, analyzing a relatively large bundle of projects to ensure that the land-use effects are significant and play a role in the final accessibility impact analysis.

The Mobility 2040 Plan

The San Antonio, Texas, metropolitan area, also known as the Alamo Area, is a fast-growing region with significant planned investment in new roadways. According to the region's long-range transportation plan, the metropolitan-area population is expected to grow from 2.0 million in 2010 to 3.4 million by 2040.[20] Employment is likewise expected to grow rapidly from 0.9 million jobs in 2010 to 1.7 million jobs. Employment centers are located in historic downtown San Antonio, as well as in a band following a circumferential highway along the northern suburbs. The metropolitan San Antonio region embodies a strong decentralization trend, with the highest population and employment growth between 2000 and 2010 occurring in suburban locations and along major highway corridors.

The entire Mobility 2040 plan is analyzed here, including all the proposed roadway expansion projects in the region's long-term transportation plan, adopted in 2014. The time horizon for the accessibility analysis is 2020, so only transportation impacts through the year 2020 are considered. The forecasts are built on the baseline year of 2010. The twenty-five-year plan includes a total of $17.2 billion in transportation funding for operations, maintenance, safety improvements, and roadway expansions, including significant investment in radial freeways extending outward from the metropolitan core. Approximately $2.1 billion of this total is applied to roadway and capacity expansions.

Forecasting Land-Use Impacts

The land-use model forecasts change in numbers of households and employment by location and type based on trends prior to the baseline year (in this case, 2005–10). The Alamo Area Metropolitan Planning Organization provided detailed land-use data for 2010, which enabled the calibration of the land-consumption component of the land-use model. Based on these data, a zone-based land-use model, known as TELUM, was used to forecast 2020 land-use patterns for both the build and no-build scenarios. Two different 2020 travel-time matrices, corresponding respectively to the build and no-build scenarios, were fed into the model to produce two distinct land-use forecasts for the future of the Alamo Area region. The challenges of land-use forecasting, together with the relatively coarse nature of the TELUM model, mean that any land-use forecast comes with a great deal of uncertainty. TELUM was used for this project not because it is the best land-use forecasting model available, but because of its ready availability to planners in local practice. TELUM's forecasts are not presented as fact, but as superior to standard travel-demand modeling with its implicit assumption of no land-use impacts whatsoever.

Work accessibility by automobile was calculated for each of the two scenarios with a gravity model (see equations 3.2 and 3.3). Because the TELUM land-use model produces population forecasts by household type and employment forecasts by industry type, a variety of accessibility metrics are possible; for example, it is possible to calculate the accessibility impacts on low-income households separately from the impacts on high-income households. However, only aggregate average population accessibility impacts are reported here. The travel cost for the accessibility calculations in this case is peak-hour auto travel times.

$$a_i = \sum_i g_j * e^{-0.14 * t_{ij}} \tag{3.2}$$

$$A_{Region} = \sum_i a_i h_i \tag{3.3}$$

A_{Region} is the total household-weighted accessibility to jobs for the San Antonio metropolitan region. It is calculated via two steps: the first step is to calculate accessibility for each zone (in this case the zones are census tracts), and the second step is to weight the accessibility in each zone by the number of households in that zone. For these equations, a_i is accessibility for zone i, h_i is the number of households residing in zone i, g_j is the total number of jobs located in zone j, t_{ij} is the auto-based travel time between zone i and j during the peak hour, and –0.14 is the impedance coefficient. Changes to mobility were also examined for each of the two scenarios by comparing the aggregate (sum total) travel times for 2020 vehicle flows across the metropolitan region. We summed projected peak-hour zone-to-zone vehicle flows adding up the peak-hour vehicular travel time for each individual vehicle flow. In equation 3.4, T is the aggregate regional sum of peak-hour travel times in minutes, f_{ij} is the flow of vehicles from zone i to zone j over the peak hour, and t_{ij} is the peak-hour travel time from zone i to zone j.

$$T = \sum_{i,j} f_{ij} * t_{ij} \tag{3.4}$$

The mobility and accessibility implications of the Mobility 2040 long-range plan are examined through a comparison of the build and no-build scenarios. The build scenario represents what would happen if all the Mobility 2040 roadway expansion projects planned through 2020 were built and the associated land-use changes occurred. The no-build scenario represents what would happen if none of the Mobility 2040 projects were built and includes the land-use patterns that would result from not introducing any new roadway infrastructure.

Anticipated household gains that result from the Mobility 2040 projects are concentrated along the northern metropolitan fringe, in particular in the northeast. Population losses are concentrated in the northern, inner-ring suburbs and much of the central city (figure 3.5). Employment gains due to the transportation plan cluster in the far northwest and northeast of the region, while employment losses are scattered throughout the central city and the northern inner ring (figure 3.6).

Table 3.2 portrays mobility and accessibility performance across four scenarios: build, no-build, speed-only effects, and year 2010. The results indicate

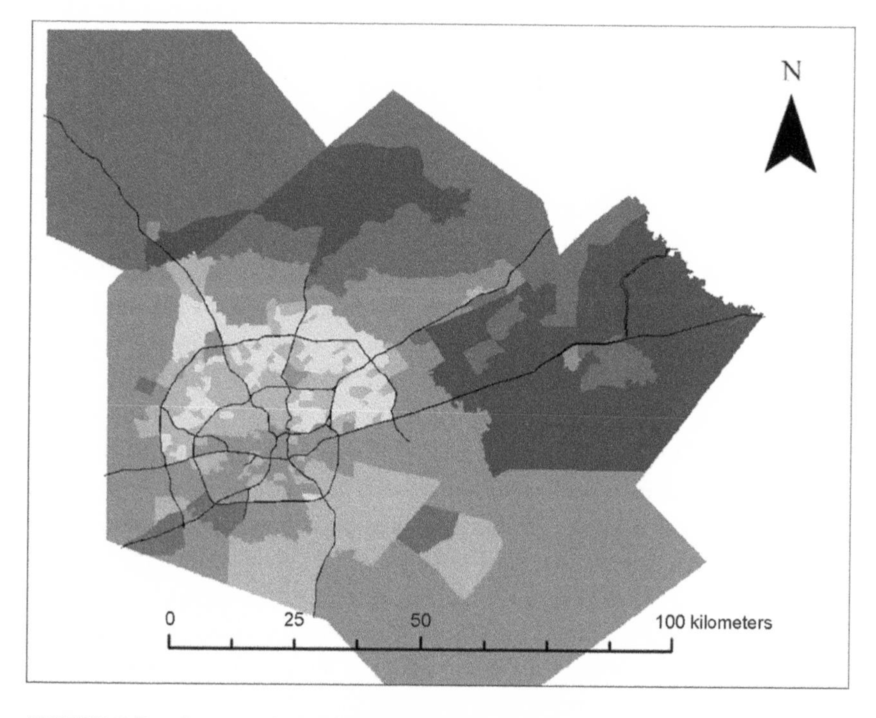

FIGURE 3.5. Forecast household shift due to Alamo transportation plan for 2020. Darker areas indicate areas with higher population growth (or less population decrease).

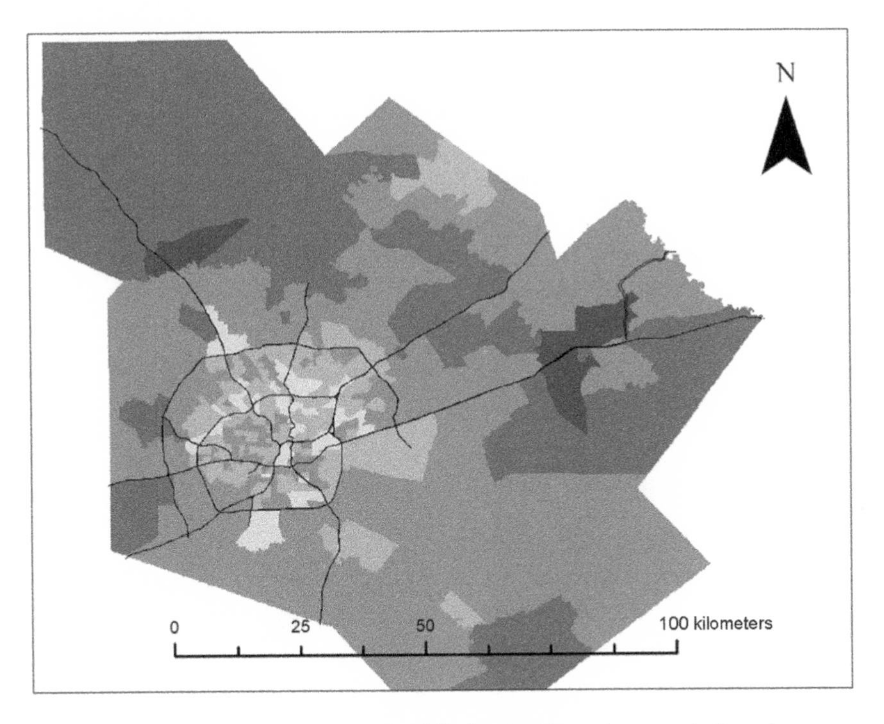

FIGURE 3.6. Forecast employment shift due to Alamo transportation plan for 2020. Darker areas indicate areas with higher employment growth (or less employment decrease).

Table 3.2 Alamo area long-range transportation plan mobility and accessibility impacts for 2020

	BUILD SCENARIO	NO-BUILD SCENARIO	SPEED-ONLY EFFECTS	YEAR 2010
Mobility impacts	Aggregate auto travel time			
Vehicular travel time, minutes	82,098,736	86,908,769	NA	73,189,146
Average trip time, minutes	24.59	26.03	NA	21.92
Percent difference no-build	−5.5%	NA	NA	NA
Percent difference with 2010	12.2%	18.7%	NA	NA
Accessibility impacts	Gravity-weighted auto accessibility to all employment			
Minimum zone	22	21	21	66
Maximum zone	96,340	96,069	97,071	107,978
Mean zone	28,105	27,790	28,298	29,330
Standard Deviation	23,335	23,635	23,553	25,382
Population-weighted accessibility	24,406	24,446	25,194	27,988
Percent difference no-build	−0.2%	NA	3.1%	NA
Percent difference with 2010	−12.8%	−12.7%	−10.0%	NA

that the mobility improvements of the Alamo's Mobility 2040 Plan through 2020 are considerable. By contrast, its accessibility impacts are negligible. The build scenario reduces aggregate peak-hour travel time by 5.5 percent. In addition, if the associated transportation projects had no land-use effect, that is, if they only changed interzonal speeds, they would increase average household accessibility by 3.1 percent. The reason the accessibility improvement is smaller than the mobility improvement is that accessibility incorporates household locations, and the speed improvements are predominantly concentrated in less-populated areas of the region. However, after the land-use effects are taken into account, the net average accessibility benefits for the region are –0.2 percent.

Mobility Analysis Is Not the Same as Accessibility Analysis

The Mobility 2040 Plan demonstrates mobility benefits through decreased average peak-hour travel times, but it offers no net accessibility benefits for the average resident of the San Antonio metropolitan region. The proposed transportation plan is forecast to decentralize both households and employment, and with these anticipated land-use shifts over a ten-year time horizon, the average household has no greater accessibility in 2020 than if no roadway expansion projects were completed at all (i.e., the no-build scenario). The advantage of faster speeds, which are documented in the mobility analysis, is completely counterbalanced by the land-use decentralization induced by the planned portfolio of roadway projects.

A contrasting case is found in a subset of the Mobility 2040 projects, termed "Loop 1604." This suite of upgrades to a ring expressway north of San Antonio was modeled for its accessibility effects compared to a baseline defined as the remainder of the Mobility 2040 projects (as opposed to a no-build alternative).[21] In this comparison, the Loop 1604 projects were forecast to concentrate land uses compared to the Mobility 2040 minus Loop 1604 forecasts, which feature extensive low-density peripheral development. As a consequence, Loop 1604's accessibility impacts were markedly positive, with the land-use impacts amplifying, rather than reversing, their mobility effects.

In the full analysis, the Loop 1604 projects offered a triply compounded accessibility benefit. First, average auto travel speeds increased, which, all else being equal, increases accessibility. Second, these travel-speed improvements were focused on concentrated areas of population and employment, so the benefit of these mobility enhancements occurred where the most people actually travel. And third, the Loop 1604 projects were forecast to spur the centralization of population and employment relative to the no-build scenario. As a consequence, a mobility-based analysis of Loop 1604 would actually significantly underestimate its auto accessibility benefits.

Although the evaluations presented here assess accessibility impacts for the average household, the method readily supports the disaggregation of accessibility impacts, including a differentiation of accessibility impacts by household type (e.g., by income), by destination type (e.g., work, nonwork, shopping, health care), and by mode (public transport, bicycle, walking). The key to the proposed method is that induced land-use effects of transportation projects must be taken into account for a full and fair accessibility evaluation.

Integrating Accessibility into Everyday Planning

Analysis of transportation or land-use projects differs from regional-scenario-based accessibility evaluation, and we have presented two methods for project-level accessibility analysis, one for land-use projects, and the other for transportation projects. Since project-based analysis must usually consider changes in either land use or transportation infrastructure, it requires tools that forecast the impact of changes in one system on the other. Furthermore, any marginal project-level change must be evaluated with respect to some meaningful basis of comparison. The methods illustrated here address these various challenges for conducting an accessibility analysis at the project level.

As we have demonstrated, transportation analysis that employs mobility tools alone can lead to perverse outcomes. Projects that enhance mobility do not necessarily enhance accessibility—indeed, they can even degrade it. The best-performing land-use development evaluated here, which is located in the central part of downtown Ann Arbor, Michigan, scores poorly when it is evaluated with traffic-impact analysis alone. The Mobility 2040 Plan, of San Antonio, Texas, improves peak-hour vehicle speeds, but actually offers no net accessibility benefits to the average household once anticipated land-use changes are taken into account.

In every case where a transportation-related planning decision must be made, there is a way to analyze that decision from an accessibility-centered point of view. If existing transportation-analysis tools consider only travel times or speeds, then these tools are mobility based, and planners who rely on such practices will fail to distinguish between accessibility-improving and accessibility-degrading projects. Realization of the accessibility shift demands new tools and methods that are applicable to decision making in everyday planning, including methods appropriate to evaluating project-level decisions.[22] This project-level analysis is an essential complement to the growing body of regional-scenario analyses that are increasingly incorporating accessibility metrics. Project-level accessibility analysis is necessary for enabling accessibility thinking to move out of the airy realm of concept, theory, and visioning and into the grounded everyday of transportation- and land-use-planning practice.

4

ACCESSIBILITY AND URBAN FORM

A prevalent view among urban planning researchers is that low-density, auto-oriented metropolitan regions are also low-accessibility areas.[1] While we argue against equating such areas with poor accessibility by definition, accessibility in these regions may in fact be measurably low. This view, if supported, would have important implications for policy reform. Compact-city and smart-growth planning approaches would have a transportation-improvement rationale. Though they are not designed for high automotive travel speeds, they could nevertheless increase accessibility and thereby yield a transportation benefit, potentially even for car travel. Transportation, environmental, and societal goals might be more synergistic than is generally thought as policy shifts allowing for closer destinations and less car travel lead to improvements in accessibility, even if they slow traffic somewhat.

Claims that higher densities degrade transportation outcomes abound, yet these are based almost entirely in a mobility logic. New York City, for instance, has been singled out as a particularly problematic transportation case based strictly on its congestion; the fact that its density offers greater origin-destination proximity than most locales has not entered into the analysis.[2] The sentiment that densities should be restricted for the sake of mobility is similarly reflected in planning conflicts from numerous countries around the world (text box 4.1).

Text box 4.1. International examples of calls for planning to limit densities based on congestion concerns

Daily Camera (Boulder, Colorado, United States). "Planning Board Rejects North Boulder Middle-Income Housing Proposal." December 2, 2016: "Just before midnight Thursday, the Boulder Planning Board rejected a proposed mixed-use development with commercial space and 50 middle-income housing units near the corner of Broadway and Iris Avenue. . . . Opponents at Thursday's hearing urged the city's Planning Board to deny the project, because it's too dense, too tall and threatens to exacerbate traffic and parking woes."

New Straits Times (Kuala Lumpur, Malaysia). "High-Density Project Gets Residents Down." July 3, 2010: "Residents of Kuchai Entrepreneurs Park want City Hall to reject a proposed high-density condominium on grounds that it will add to the congestion in the area. The residents learnt of the development through a sign erected at Lot 1221, Jalan Kuchai Lama about two weeks ago."

Sentinel (Stoke-on-Trent, United Kingdom). "900-Home Plan Is Bunkered." September 1, 2016: "'INAPPROPRIATE' plans to build 900 homes on a former golf course have been blocked. . . . Haddon Property Developments wanted to build the houses—along with shops and a school—on the site of Gorstyhill Golf Course, near Crewe. Villagers in Wychwood, which is encircled by the course, objected to the scheme, saying it would mean the loss of a leisure facility and lead to traffic chaos."

Prince George Citizen (Prince George, British Columbia, Canada). "Council Rejects Housing Project." April 26, 2017: "City council has rejected a rezoning application for a multifamily residential development in a Spruce Street neighbourhood. . . . Coun. Brian Skakun opposed the application.

"'This makes no sense to me. I support the residents. This is making a bad problem worse especially in terms of traffic,' Skakun said."

While higher densities frequently degrade traffic speeds, they could enhance accessibility if the increase in proximity they support outweighs any slowing of travel (see chapter 3). If this were the case, the densities would confer a transportation benefit, rather than creating a problem to be solved. And the converse would also be true: land-use regulations to enforce a low-density development pattern could be potentially harmful to transportation goals (defined again in accessibility terms) even if they succeeded in keeping the traffic flowing rapidly. A nationwide comparison of metropolitan job accessibility by auto is suggestive in this regard: of the ten metropolitan areas with the greatest job accessibility by auto, six are also found on the list of the ten areas with the greatest *loss* in job accessibility due to congestion.[3] While congestion reduces accessibility, it appears that some external factor tends to lead both to high automotive accessibility and to high roadway congestion. This factor may be urban form. If one defines accessibility in urbanist terms—an approach we reject (see chapter 2)—then the relationship of land-use patterns to accessibility is a foregone conclusion. But low-density regions with a high degree of separation of land uses do not necessarily demonstrate low accessibility. This is because a low-density pattern of settlement can have high accessibility (at least by motorized modes) if speeds are fast enough to compensate for the great distances. Which urban forms lead to higher accessibility is an empirical, not a definitional, question.

Empirical analysis of the relationship between urban form and accessibility can be accomplished by comparing the same regions over time or by comparing a range of regions at a single point in time.[4] In this chapter, we compare accessibility across metropolitan regions in the United States using this cross-sectional approach, estimating auto-based work accessibility metrics for thirty-eight of the largest fifty metropolitan regions in the United States. The automobile focus here is designed to test for potential accessibility benefits to compact metropolitan urban form that are not self-evident. Compactness supports public-transport accessibility (see chapter 5) in line with expectations; by contrast, low-density development is often thought to benefit auto-oriented transportation systems.

Much of the prior research on measuring accessibility is focused on case studies of single metropolitan regions.[5] A smaller body of work has compared accessibility between regions.[6] The approach presented here can be used to analyze accessibility differences across regions as well as accessibility change over time within and between regions, and can help assess the factors enhancing or degrading regional accessibility, including metropolitan compactness.[7]

Accessibility Comparison between Regions

Intermetropolitan comparisons of accessibility present a challenge because people in different regions have different propensities to travel in response to travel time or cost. The concept of "minute," "work destination," and "nonwork destination" can be reasonably consistently defined between regions. By contrast, people's willingness to take trips, a key element of virtually all measures of accessibility, varies markedly from one region to another. A traveler in Los Angeles, for example, is more likely to make a trip of sixty minutes than a traveler in a smaller region like Omaha. Whereas in some regions, such a trip would be considered long and hence avoided if at all possible, in others it would be routine. This is at least in part a function of varying urban patterns; if there are few destinations nearby, but many farther away, people are inclined to travel relatively long distances. If the reverse is true, and there are many destinations nearby and few far away, people find much less need for long trips. There may also be interregional differences that are unexplained by the physical environment in people's innate propensity to travel.

In gravity-based models of accessibility, differing propensities to travel across regions are expressed in differences in coefficients of the impedance function (see chapter 3), with higher values indicating a lower willingness to travel. The phenomenon has its expression in other metrics of accessibility as well. For example, the cumulative-opportunities model defines a travel-time radius, for example, 15, 30, or 45 minutes. Choice among these radii is analogous to the choice of an impedance coefficient, with a short travel-time radius equating to a high coefficient. The choice of both impedance coefficients in gravity-based accessibility and the search radius in the cumulative-opportunities model will influence both accessibility metrics, and, potentially, the ordinal accessibility ranking of the metropolitan regions in the study.

In the customary application of the gravity model, the impedance coefficient is estimated separately for each region. Use of these varying functions for the comparison of accessibility across regions would lead to a logical inconsistency, however. Regions whose land-use and transportation systems require a lot of time and money in travel demonstrate a higher propensity to travel further. For this reason, using region-specific impedance coefficients in intermetropolitan comparisons of accessibility is tantamount to giving the region with harder-to-reach destinations "credit" that it does not deserve. A consistent impedance coefficient, by contrast, creates an evenhanded metric on the ability to reach destinations across regions; it treats a minute as equally burdensome no matter where a person lives. Therefore, the impedance coefficient used here is the expected value for the parameter in the median region in terms of population.[8]

Evaluating the Interaction of Speed and Proximity

In order to explore the interaction of speed and proximity on accessibility, metropolitan areas were paired on the basis of population size, and the distribution of accessibility was analyzed between the two regions. For example, a pairing of metropolitan Washington, DC, with the San Francisco Bay Area reveals similar levels of accessibility at the low end (e.g., the 1st-percentile household in accessibility terms) and the high end (e.g., the 99th-percentile household in accessibility terms). The rest of the distribution reveals higher accessibility for the San Francisco Bay Area (figure 4.1). Though the horizontal axis in these graphs is ordered simply by population percentile of the accessibility score, it has something of a geographical interpretation: since accessibility generally declines as one moves outward from the center of the region, households at the low end of the distribution tend to reside in peripheral areas, while those at the high end live at the center.

Accessibility differences between two regions may be decomposed into a proximity component and a speed component.[9] This is accomplished by transforming the speed distribution of San Francisco into that of Washington, DC.

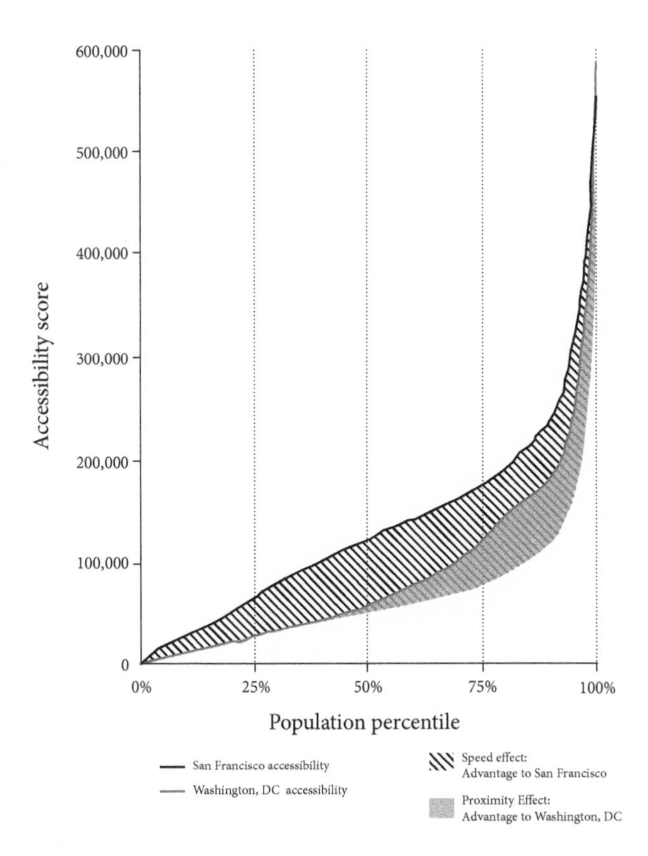

FIGURE 4.1. Decomposition of accessibility differences between metropolitan San Francisco and Washington, DC

A new set of accessibility indicators are calculated for San Francisco, using travel times derived from Washington speeds, generating a new accessibility curve (figure 4.1).[10] The transformed curve, which in this case is the lowest of the three, represents the accessibility distribution that would hold for the San Francisco Bay Area if travel speeds there matched those of Washington, DC. For this reason, the gap between that curve and the San Francisco accessibility curve is entirely speed related, while the difference with the Washington, DC, curve is entirely proximity related. The speed-related advantage to San Francisco is thus shown as the cross-hatched area between the top and bottom curves; the proximity-related advantage to Washington is represented by the shaded area below the bottom curve. Notwithstanding the greater density of the San Francisco Bay Area, Washington, DC, demonstrates a proximity advantage. But that advantage is negated by the greater magnitude of Washington, DC's speed disadvantage; the potential accessibility benefit of greater proximity was squandered by poor mobility relative to that of the San Francisco Bay Area—in this case, automobility, since the accessibility metric is automobile-based. Here, the accessibility outcome is consistent with traditional mobility-based transportation planning; poor mobility has degraded the accessibility of what might otherwise be a highly accessible metropolitan area. At the same time, this comparison demonstrates the redundancy of defining twin goals for "mobility and accessibility." Where mobility is sufficient to overcome a proximity disadvantage—as in the San Francisco case—the outcome materializes as superior accessibility. Because accessibility incorporates mobility, mobility does not need a separate accounting to assess its benefit.

The relationship between speeds and accessibility demonstrated in the Washington–San Francisco comparison is not universal, however, as illustrated by a similar decomposition of accessibility differences between another pair of similarly sized metropolitan areas: Philadelphia and Houston (figure 4.2). Philadelphia enjoys a considerable accessibility advantage over Houston for most of the population distribution. A decomposition of accessibility between the two regions reveals that Houston enjoys a large speed advantage over Philadelphia, but suffers from a proximity disadvantage. Notably, the proximity disadvantage exceeds the speed advantage, generating an accessibility disadvantage for Houston overall. While the Washington–San Francisco comparison illustrates how mobility can create a higher accessibility outcome, Houston-Philadelphia provides a counterexample: Houston's accessibility suffers when compared to Philadelphia's despite Houston's faster travel speeds. The vast distances in low-density Houston render it less accessible than the slower but more compact Philadelphia.

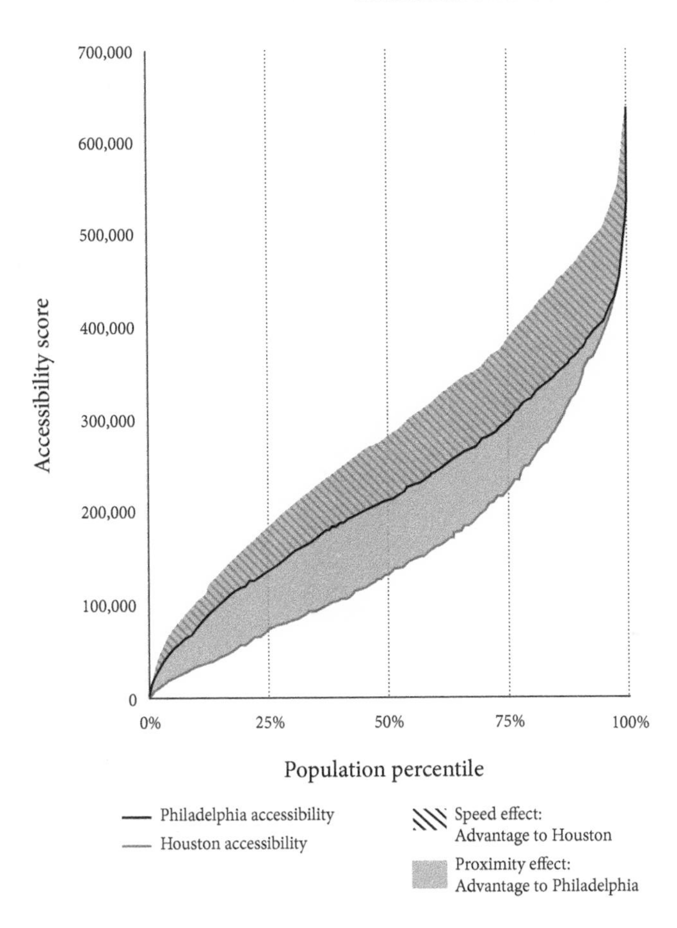

FIGURE 4.2. Decomposition of accessibility differences between metropolitan Philadelphia and Houston

A third case presents itself as well: if region "A" enjoys both a speed and a proximity advantage over region "B," it will demonstrate higher accessibility overall. This is the case with New York as compared with Los Angeles (figure 4.3). New York enjoys a slight speed advantage, a considerable proximity advantage, and an overall accessibility advantage over Los Angeles for most of the population distribution. This finding puts the view of New York as a particularly problematic transportation case in a different light.[11] Notwithstanding the serious congestion problems of New York City, its region has the highest accessibility of all regions studied (a function in part of its very large size).

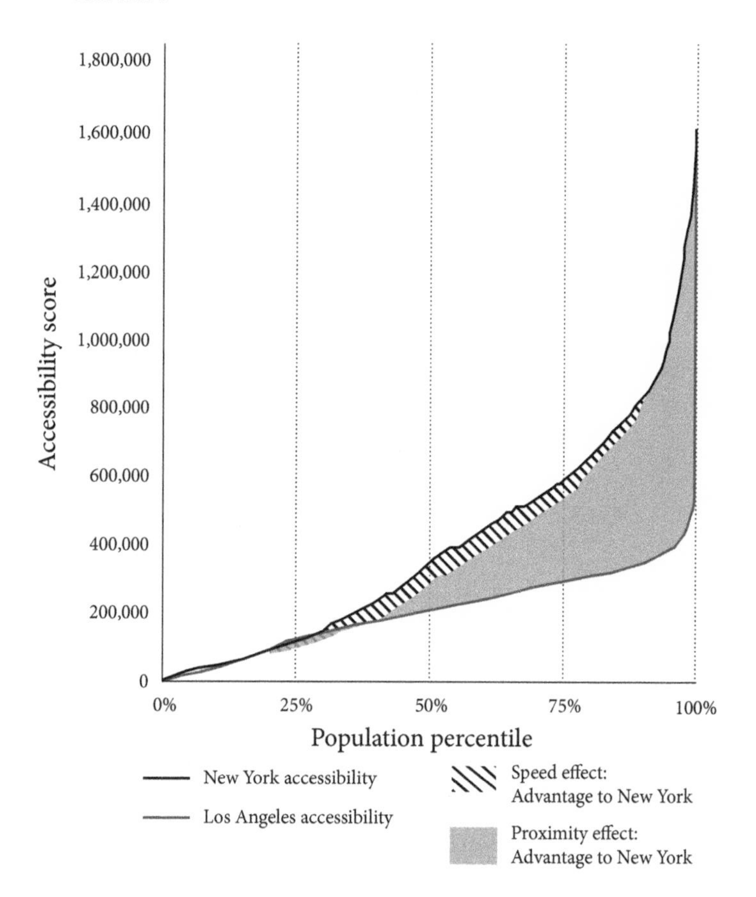

FIGURE 4.3. Decomposition of accessibility differences between metropolitan New York and Los Angeles

Accessibility and Urban Form Study

The analyses we have presented focus on metropolitan size but do not account for urban form. To develop data sufficient to test the accessibility/urban-form relationship, metropolitan planning organizations (MPOs) of the fifty largest U.S. metropolitan regions (ranked on the metropolitan statistical area [MSA] population count of the year 2000) were contacted with a request for travel-demand modeling data, the central piece of information needed to calculate peak-period auto- and employment-based accessibility scores. Due to variations in travel-demand modeling data across different metropolitan regions, sufficient data from thirty-eight regions were received; the remaining twelve regions were left out of the study (appendix A, table 4a).

The most important data items are matrices of interactions between all zones in a metropolitan region, including travel times and the number of trips between zones. Local definitions of the peak period differed between regions; in each case, the definition of the morning peak period in use in the area was employed. Models for generating travel times varied between metropolitan areas, and some of the intermetropolitan variability in data would be attributable to these differences in data collection and modeling methods, introducing error of unknown magnitude into future analyses.

For work and nonwork destinations, data on business establishments were purchased from the private vendor Claritas, Inc.[12] This dataset contains several attributes of business establishments, including the number of jobs at each location in 2008, and classification codes from the North American Industry Classification System (NAICS), allowing identification of businesses by industry type. All business establishments within the thirty-eight regions were geocoded to the street-address level, then aggregated by traffic-analysis zones into the number of work and nonwork destinations by zone.

This study uses the gravity model as a "place-based" measure. All people residing in a zone are assigned the same level of accessibility, regardless of their personal choices of destinations or preferences for travel. Because each metropolitan area experiences a range of accessibility distributed across its population, the accessibility experienced by the median resident was selected as a summary statistic for evaluating overall regional performance.

Density's Relationship with Accessibility

An examination of the relationship between accessibility and urban form context could entail a host of metropolitan characteristics, including centralization, concentration, density, and others.[13] However, exploratory regression models revealed little to no relationship between such metrics as centralization or concentration and accessibility; by contrast, average regional densities appeared to be a significant determinant of median work accessibility by automobile.

The finding that several common urban form indicators have little relationship to regional accessibility is, in part, a function of a focus on the median resident. The median resident in accessibility terms of the typical U.S. metropolis is a suburbanite (in each of the thirty-eight regions, the zone with the median accessibility score is located outside the central city, and in many instances at a substantial distance from the central city). In nearly every case, this individual does not live in or near the downtown, or even in or near a suburban concentration such as a downtown or transit-oriented development. Thus, the extent of these concentrations affects this person only marginally. By contrast, overall regional densities can affect median accessibility markedly in two ways: speed and proximity.

SPEED

Higher densities can reduce average travel speed. Auto-ownership rates in U.S. metropolitan regions, including higher-density regions, are high. Thus, population density in these regions can lead to high traffic densities and therefore slow speeds. If distances are held constant, slower travel speeds will degrade accessibility.

PROXIMITY

Higher densities can increase proximity by shortening the distance between origins and destinations. Higher-density regions put numerous destinations closer to a given origin than their lower-density counterparts. For example, in the case of jobs, higher job densities mean more job locations are within a closer proximity, increasing accessibility.

Thus, the effect of density on accessibility can be thought of as the sum of the speed effect and the proximity effect. If the speed effect dominates, low-density regions will be more accessible; if the proximity effect dominates, high-density regions will be more accessible.

There are good reasons to believe that the speed effect might dominate, a phenomenon that would render low-density regions more auto accessible than their higher-density counterparts. To illustrate, consider the cumulative-opportunities measure, or the number of destinations reachable within a given amount of time. The territory accessible within Y minutes is an irregularly shaped area (depending on the shape of the street network) but can be roughly conceived of as a circle (appendix A, figure 4a). When speed doubles, the radius of the circle that can be reached within a given time doubles as well. As a consequence, the reachable area within the circle quadruples. Given a constant density of destinations, those places that can be reached within the specified time also quadruple with the doubling of speeds. In this stylized case, accessibility increases with the square of speed.

The impact of increasing densities on accessibility can be illustrated in a similar fashion. If speeds are held constant, but the density of destinations doubles, accessibility doubles (appendix A, figure 4b). Thus, while accessibility increases with the square of speed, it increases linearly with density. For this reason, increasing speed confers a very significant accessibility advantage, one that will be difficult to overcome with the proximity effect. This observation tends to support the mobility-centered view that is reflected so dominantly in current transportation policy—but there is another possible explanation for density's influence on accessibility.

Density and Speeds

Metropolitan density may influence auto travel speeds in two competing ways. First, low-density regions typically offer more roadway capacity per person than their high-density counterparts. Low-density regions in the United States tend to have a high ratio of roadway lane-miles per capita, a factor that would tend to raise travel speeds (figure 4.4). Yet the tendency for higher roadway capacity in low-density regions appears to be at least partly negated by the higher vehicle-miles traveled (VMT) per capita observed in these regions (figure 4.5). Low-density regions are thus simultaneously roadway-intensive and travel-intensive. Speeds are determined neither by vehicle-miles traveled nor by roadway miles in isolation, but rather as a function of the interaction of the two. One broad indicator of this interaction is traffic density, measured as the ratio of VMT to freeway lane miles. The relatively strong (r^2=0.26) negative relationship between density and VMT per capita interacts with a somewhat stronger (r^2=0.37) relationship between density and freeway lane miles per capita. The net result is that the relationship between population density and traffic density, while positive, is relatively weak (r^2=0.11) (figure 4.6). Lower-density regions show less traffic density than their higher-density counterparts, but only marginally. Thus, while the speed-accessibility link

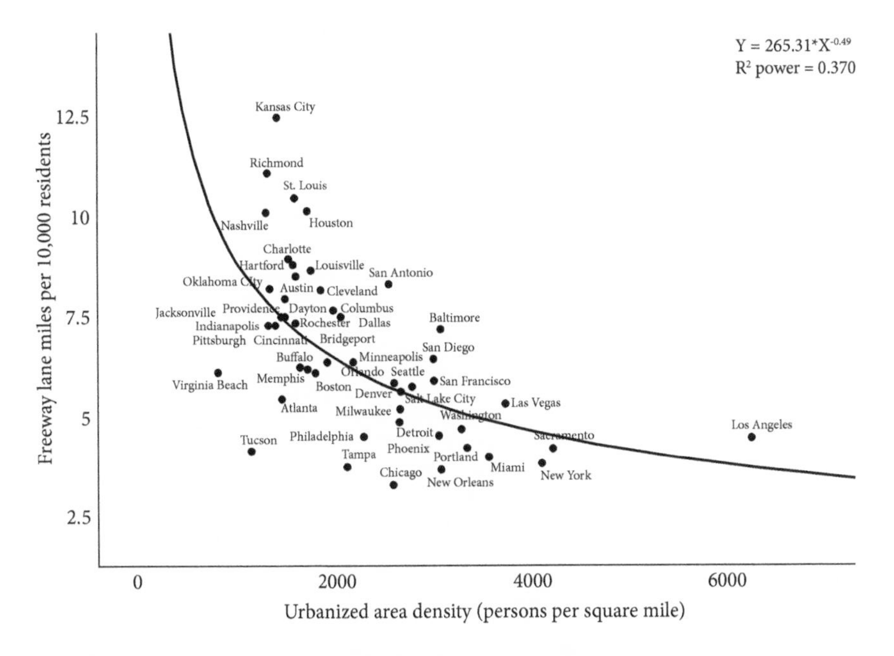

FIGURE 4.4. Urbanized area density and freeway lane miles per 10,000 residents. U.S. Department of Transportation, *Highway Statistics 2008* (Washington, DC: Federal Highway Administration, 2008).

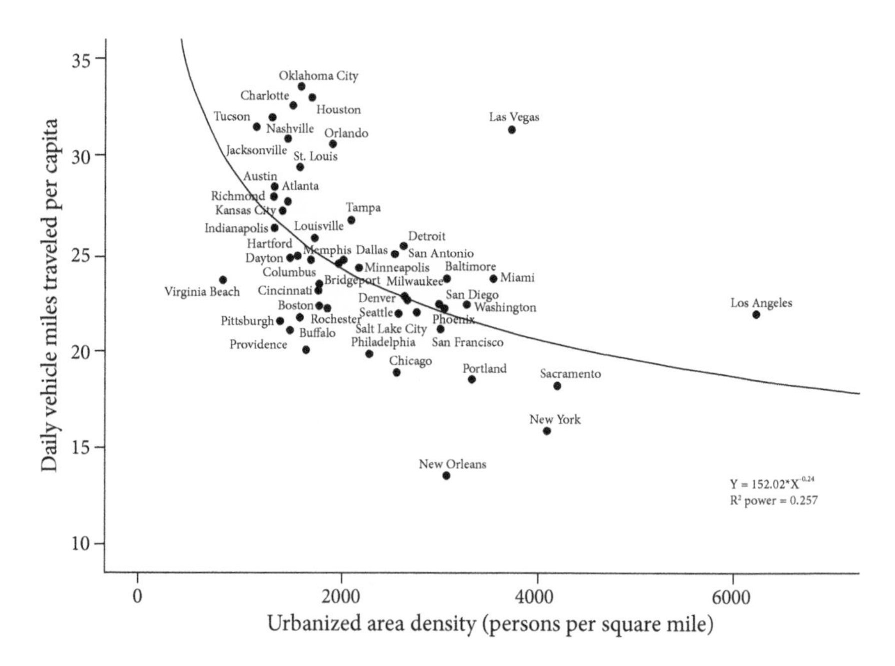

FIGURE 4.5. Daily vehicle miles traveled per capita by urbanized area density, largest fifty U.S. (census-defined) urbanized areas. U.S. Department of Transportation, *Highway Statistics 2008* (Washington, DC: Federal Highway Administration, 2008).

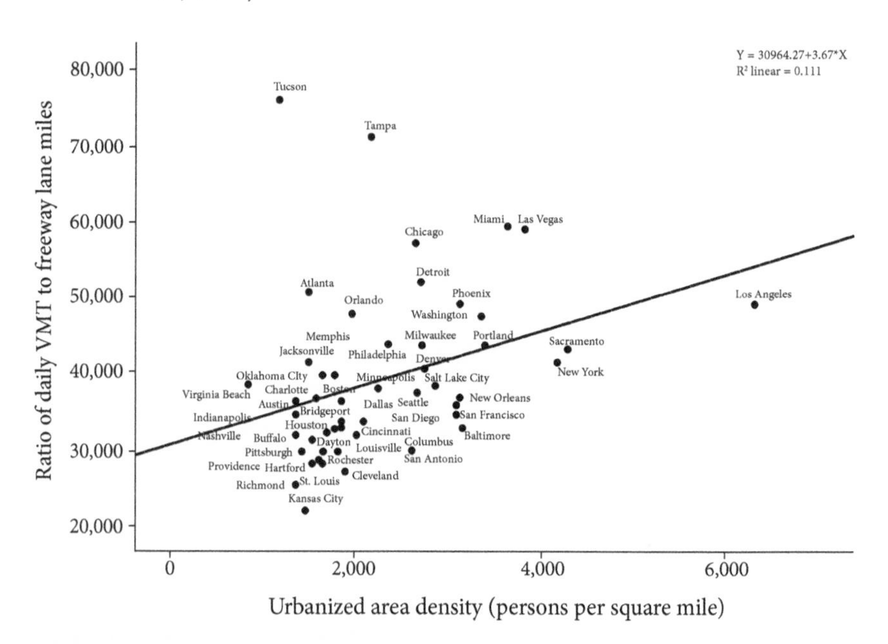

FIGURE 4.6. Traffic density by population density

Source: U.S. Department of Transportation, *Highway Statistics 2008* (Washington, DC: Federal Highway Administration, 2008)

is expected to be strong, the density-speed link appears to be quite weak and, by extension, may lead to a weak link between density and accessibility.

These relationships were tested through a path analysis, using the thirty-eight U.S. metropolitan areas for which data were available. Path analysis is an application of multiple regression that aims to identify dependencies among a set of variables. It uses several regression equations in a recursive manner to estimate a system of interrelated variables, and it is customary to represent the results graphically as shown in appendix A, figure 4c.

The analysis tests the strength of relationships along causal paths. In this analysis, the effect of density on accessibility via speed is compared to the effect of density on accessibility via proximity. As expected, the effect via speed is negative and the effect via proximity is positive. But most important are the relative magnitudes: notwithstanding the conceptual advantages of speed in generating accessibility, density exerts a positive accessibility effect via proximity that is much stronger than its negative effect via speeds. This finding is consistent with a 2018 study demonstrating that proximity exerts a stronger influence on job accessibility than speeds.[14]

These results—with the positive impacts of density on auto accessibility outweighing its negative impacts—are corroborated by the data displayed in figure 4.7. Overall, the figure demonstrates a positive relationship between urbanized-area density and accessibility. There is also some correlation between density and metropolitan size—New York and Los Angeles are simultaneously two of the largest and densest regions—but the positive relationship holds even without these cases. And even the small region of Las Vegas demonstrates high accessibility, in part as a function of its population density.

Las Vegas represents an interesting case for another reason as well. It is generally considered to be a fairly sprawling metropolitan area; for example, Reid Ewing, Rolf Pendall, and Don Chen ranked it as the 48th most sprawled region of 83 analyzed.[15] Indeed, in the dimension of neighborhood mix of homes, jobs, and services, Las Vegas was much more sprawling than the typical region. Yet its high levels of automobile accessibility illustrate the problem with defining accessibility as sprawl's antithesis: a sprawling region can offer high auto accessibility. While Las Vegas has achieved this through density, other regions have achieved a similar outcome differently. For example, Kansas City, a low-density region that is the 29th most sprawled in the study, nonetheless achieved high levels of auto accessibility because it is relatively free from traffic congestion.[16] Public-transport accessibility is another issue entirely, and it is quite difficult for a low-density region to achieve high scores in that regard. For example, Kansas City's public-transport accessibility ranks 38th out of the 49 metropolitan regions studied by Andrew Owen, Brendan Murphy, and David Levinson.[17]

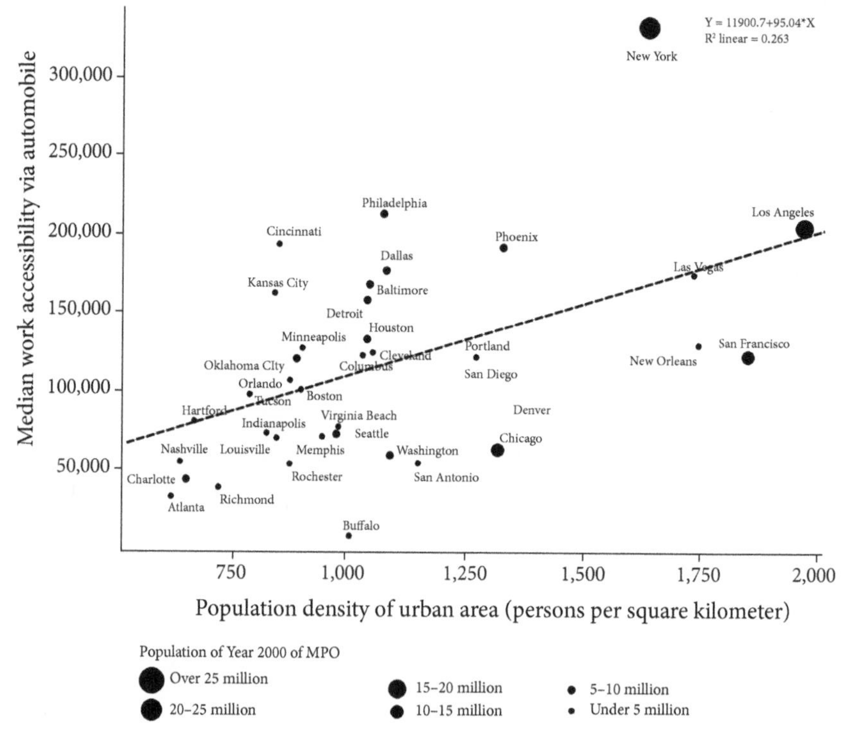

FIGURE 4.7. Median work accessibility via automobile by urbanized area density

Metropolitan Densities and Transportation-Service Quality

For metropolitan areas in the U.S. overall, there is a positive relationship between density and auto accessibility. This suggests that land-use policy can be highly relevant to accessibility outcomes. In particular, allowing metropolitan compactness by easing land-use regulations mandating low development densities can improve the effectiveness of the roadway transportation system as defined in accessibility terms. Yet the success of the strategy remains a matter for empirical accessibility evaluation. The relationship between metropolitan density and work accessibility via automobile, while positive, is hardly ironclad, and some low-density regions do offer high auto accessibility. And transportation success overall should not just consider the automobile but should consider accessibility via all travel modes, including walking, cycling, public transport, and cars.

By contrast, traditional mobility-based transportation evaluations tend to militate against denser development on the theory that dense land use can lead to dense traffic and hence congestion. The analysis presented here suggests a more nuanced understanding of dense development: increasing density may well be accessibility enhancing if the proximity effect on accessibility outweighs the speed-reduction effect. Conversely, land-use regulations that constrain such densification may degrade accessibility by car even as they strive to maintain or enhance mobility by automobile. Whereas higher densities are frequently viewed as a transportation disadvantage because of their impacts on automobility, they may offer distinct transportation advantages when viewed in accessibility terms.

5

THE SPECIAL CASE OF PUBLIC-TRANSPORT ACCESSIBILITY

Public transport plays a special role in accessibility. Like driving, biking, or walking, its first job is to provide the mobility that allows travelers to reach their destinations. But public transport differs from other modes in its capacity to boost proximity indirectly; it can bring origins and destinations closer together to shape land-use development toward higher proximity. "If a new subway station opens near your house, you get improved mobility," writes public-transport consultant Jarrett Walker. "But if the station leads to new development around it, you may get a new grocery store close to home. Even if you never ride transit, that's an access improvement for you."[1]

These proximity impacts operate at both local and regional scales. When public transport enables large and concentrated downtowns to thrive, or when it facilitates metropolitan compactness overall, it is supporting a regional accessibility benefit that accrues to transit users and nonusers alike. Other modes can influence proximity as well. But investments in automobile infrastructure frequently lead to proximity reductions through the spreading of land-use development, an effect that can degrade or even reverse the investments' accessibility benefits (chapter 3).[2] And neither cycling nor walking generally shape the proximity of metropolitan development with the strength and magnitude of public transport.

Public transport also differs from other modes in that measuring its accessibility involves a series of technical challenges that planners and public officials must confront when they seek to extend the accessibility shift to the transit context. In this chapter, we consider the special challenges of public-transport accessibility measurement and then demonstrate one application of such measurement: the assessment of bus versus rail transportation deployments for their respective contributions to metropolitan accessibility.

Technical Challenges in Measuring Public-Transport Accessibility

Standard approaches to accessibility measurement require adaptation in order to account for several challenges particular to the public-transport mode. The first and most fundamental conceptual challenge is defining travel time itself. A public-transport trip from an origin to a destination typically involves distinct phases: access time (the time to walk, cycle, or drive to a transit station to board the service), waiting time (the time between a traveler's arrival at a station and the vehicle's arrival at the station), boarding time (the time for the vehicle operator to collect fares from passengers), in-vehicle time (the time a vehicle is moving from one station to the next), transfer time (the time to transfer from one line or mode to another), and egress time (the time to travel from the station to the final destination). Although in-vehicle time is the most important, all elements would ideally be included in the total public-transport travel time. But regional planning agencies have defined their overall public-transport travel time with varying mixes of the other elements, making comparisons of accessibility difficult across metropolitan regions or across modes for lack of consistent definitions.[3] While many of these elements of travel time are at least in principle present in other modes, their role in public transport is much greater than in its competitors.

Second, in contrast to automobile drivers, cyclists, and pedestrians, who rarely have to wait before they can begin a trip, public-transport riders are constrained by a service schedule. The periodic nature of public-transport service is a challenge to accessibility measurement. Walker illustrates the importance of vehicle-arrival frequency for public-transport riders with an analogy to automobile drivers: "Imagine a gate at the end of your driveway that opens only once every 30 minutes."[4] Service that runs every 15 minutes is far more valuable to a public-transport patron than service that runs every 30 minutes. Headways, the gap in minutes between consecutive vehicles at a stop, tend to vary substantially by location within a metropolitan region, often ranging from frequencies as small as 10 minutes or less on high-traffic corridors in the urban core to frequencies as large as 60 minutes or more for lines in outlying suburban districts. Frequency also varies by time of day, so that a line that runs every 30 minutes during the day might shift to 60 minutes during evening hours, and then discontinue service altogether late at night. Frequency varies by city size as well.

Small cities commonly run lines at headways greater than 60 minutes throughout the system.[5] Accounting for these differences in frequency is essential to making comparisons of accessibility within and across metropolitan areas.

A common approach to accounting for the spatial and temporal variation in service frequency in accessibility calculations has been to ignore the frequency of public-transport lines. Such a procedure makes the simplifying assumptions that public-transport trips are taken at the optimal time, and that travelers are familiar with the schedule and deliberately time their trip to meet the bus or train at its scheduled arrival time.[6] For riders where service is scheduled with long headways of, for example, 30 to 60 minutes, such an assumption may reasonably represent their behavior because these travelers can expect long waits if they arrive at a station randomly. Even if the assumption reflects most actual behavior, it fails to account for the inconvenience associated with adherence to a schedule, especially if an analyst wants to compare the accessibility of public transport to that of other modes like auto and bicycle, which do not face similar schedule constraints. For riders whose service is scheduled with short headways, the assumption does not likely match actual behavior and thus undercounts the total travel time by ignoring wait time.

Early studies that tried to get around this limiting assumption applied a weighting factor to accessibility metrics to account for differences in frequency, giving more accessibility credit to public-transport stations with higher-frequency service on average throughout a day.[7] Because public-transport travel time between an origin and destination is highly dependent on departure time, some studies after 2010 took a more complex approach by generating a travel time for multiple departure times—in some cases as often as every minute—in a peak period for any single origin-destination pair, and then estimating an average travel time for that period.[8] Although using such an approach provides for a more accurate description of public-transport accessibility, a simplified approach of using a single, optimal departure time can provide a reasonable approximation in many cases.

A third challenge for public-transport accessibility is posed by transfers to and from other modes. Every public-transport trip involves a mode other than public transport at the origin and destination, usually by walking but sometimes by other modes such as bicycle or automobile at park-and-ride stations. Strictly speaking, auto accessibility does as well. But auto accessibility is commonly calculated without considering the walking time at the origin and destination because walking time makes up a negligible share of the total travel time for most trips. Given the prevalence of parking spaces and a built environment oriented toward serving auto trips, this is a sensible assumption for auto accessibility, at least in the United States. But public-transport accessibility must include these elements of the total travel time. One approach is to assume an average walking speed applied to a network distance from an origin zone's centroid to the starting public-transport stop, and from the ending station to the destination zone's centroid. An analyst might choose to set a threshold walking distance and eliminate any zones that fall beyond the threshold. To illustrate, any zone with a centroid farther than two kilometers from a public-transport station might be assumed to be outside of transit's service area and assigned a value of zero for transit accessibility. Another approach is to avoid eliminating any zones on the assumption that any zone can reach a public-transport stop, even if it requires the highly improbable walking distance of several kilometers. In this case, every zone in a metropolitan space would be assigned a nonzero value for public-transport accessibility, with zones at a great distance from stations having miniscule values.

Fourth, public transport itself typically consists of several variations of the mode, including technologies such as commuter rail, heavy rail, light rail, streetcars, buses, and express bus service. These technologies are intended to offer highly varying types of service that tend to cater to distinct passenger groups, with spatial and temporal variation. Therefore, a single public-transport accessibility value may not be a meaningful metric. In some cases, an analysis of accessibility might be more appropriately subdivided into categories of, for example, rail accessibility and bus accessibility. In others, a weighted hybrid of the various public-transport technologies may be appropriate.[9]

A fifth challenge is that the inherently multimodal nature of public transport requires modeled networks that tend to be less developed than the modeled networks used for calculating auto accessibility. Zone-to-zone travel times for automobile travel are modeled using a detailed road network that has typically been developed and regularly maintained over many years. By contrast, many local and regional public-transport agencies lack the capacity to maintain consistently updated network data on routes and service schedules.[10] Furthermore, while a road network is typically constructed and maintained by a regional or state level of government, public-transport services in most large metropolitan regions are provided by multiple separate agencies, which can introduce inconsistency in the regional network. However, the widely used General Transit Feed Specification (GTFS) holds promise for addressing this inconsistency by providing a standardized format for updating and disseminating network schedules and geospatial data.[11] Still, pedestrian and bicycle networks—required for calculating the time to arrive and depart from a public-transport trip—remain highly undeveloped and uneven. Analysts often assume that walking and cycling can follow the paths of the road network despite the large share of roadways in most regions that offer poor or nonexistent walking conditions. Pedestrian and bicycle paths that do not appear in road networks are frequently not taken into account.

Finally, unlike travel by automobile, travel by public transport regularly involves transfers from one line to another or even from one transit system to another. Accommodating transfers in the calculation of public-transport accessibility is a moderately complex technical challenge that often requires computer programming skills and a high level of computer resources. Analysts might assume a limit of one or two transfers, a reasonable assumption given that few patrons transfer more than twice on a trip, because the computational complexity of shortest-path algorithms increases dramatically with the number of transfers.[12]

Misconceptions about Public-Transport Accessibility

Despite the special role that public transport plays in accessibility, planners traditionally define success in public-transport systems strictly in mobility terms, and often in ways that distort or muddle the concept of accessibility. Standard metrics used in practice assume that the outcome of public transport that matters is moving vehicles or people; they fail to address how well public transport is doing its job of connecting people with their destinations. For example, the *Transit Capacity and Quality of Service Manual*, a guidebook produced by the U.S. federal government to assist local agencies in evaluating service quality, recommends a range of performance metrics without any reference to accessibility as defined here.[13] Not a single evaluation measure in the widely used guidebook can help policy makers decide whether their public-transport system performs well in helping patrons reach their destinations. Examples of the most important metrics include service frequency (average time interval of vehicles arriving at a stop), hours of service (span of time that service is available throughout a day), and service access (share of population that lives within walking distance of a public-transport stop), all of which have their purposes, but all of which assess mobility alone.

This last metric, service access, is often confused with accessibility.[14] For example, a report from Arlington County, Virginia, identifies in a section titled "Performance Measures for Accessibility" a metric defined in the report as "Percent of population and employment in close proximity to pedestrian, bicycle, car share, bike share, and transit facilities."[15] The report asserts that this metric helps gauge the capacity to reach destinations: "Proximity to multi-modal transportation services indicates the ability for a resident or employee to be able to access Arlington's and the region's destinations, services and activity centers using the most appropriate travel mode for their trip."[16]

Distance to a transit station is not a measure of accessibility to destinations, but rather an indicator of access to the public-transport system. It indicates people's ability to move via transit rather than their ability to reach destinations; it is a mobility indicator and not an accessibility indicator. Distance to a station addresses only the conditions at the origin of a trip without considering the linkage between origin and destination. A neighborhood may have an exceptionally high degree of access to the public-transport system, with many residents living near a station, but if reaching a destination from that station requires a lengthy, circuitous route interrupted by several transfers, this would not correspond with high accessibility.

Both the passenger volumes that a public-transport system experiences and the concept of accessibility are needed for planning and evaluation, but they must remain conceptually separate since high accessibility may accompany low usage, and vice versa. For example, a wealthy neighborhood may boast a high level of public-transport service that offers excellent accessibility, but because most residents own cars and prefer to drive them, few ride public transport. By contrast, a low-income neighborhood with many residents who have little choice but to travel by public transport may have high ridership even if the neighborhood's public-transport service yields poor accessibility. Public-transport services aspire to both high passenger usage and high-quality service; since accessibility is the service people seek in their transportation systems, accessibility metrics are needed to gauge the service dimension.

The technical complexity associated with the accessibility shift partly explains why planners and policy makers rely heavily on mobility-based evaluation in public transport, but this shift is nevertheless essential if public agencies intend to improve the accessibility their patrons seek when they travel. Although accessibility measurement is more challenging for public transport than for other modes, the view that a better public-transport system is one that offers people the opportunity to reach more destinations for a given investment of time and money provides a measurable basis of progress toward desirable land-use and transportation outcomes. The rest of the chapter uses an empirical study to demonstrate one use of accessibility metrics to evaluate the kinds of public-transport technologies and land-use developments that can lead to better public-transport accessibility in metropolitan areas.

What Matters Most When Planning for Public-Transport Accessibility?

The choice among appropriate public-transport technologies has divided planners for decades, pitting bus and bus rapid transit (BRT) investments on the one hand against rail (frequently light rail transit or LRT) on the other. These public-transport technologies have been evaluated and compared on numerous dimensions, including cost, impact on property values, capacity to affect physical development, attractiveness to riders, capacity to offer congestion relief, and role in urban revitalization and economic productivity.[17] Urban revitalization, enhanced economic productivity, and highway congestion relief may be desirable by-products of public-transport investment, but none are as fundamental as accessibility enhancement.[18]

Given the centrality of accessibility to transportation planning, one might expect to find comparisons of bus and rail transport in terms of their respective contribution to people's capacity to reach destinations. Such studies are surprisingly rare. One study evaluated accessibility change with an LRT deployment, but this research was on LRT alone and was not designed to compare bus and light rail.[19] A second study modeled the impact of the size of a public-transport system on productivity (defined in accessibility terms) and in the process presented models of bus and rail contribution to metropolitan work accessibility.[20]

The analysis presented here uses the same dataset as this second study: public-transport accessibility data from the University of Minnesota Accessibility Observatory.[21] Building on previous research on the components of auto accessibility across multiple metropolitan regions, the analysis presented here assumes that public-transport accessibility is a product of the interaction of the mobility provided by transportation modes and the proximity provided by metropolitan land-use patterns.[22] Metropolitan public-transport accessibility is modeled here as an outcome that depends on factors including the population of the metropolitan region (larger regions tend toward greater accessibility), its urban form (denser concentrations are expected to boost public-transport accessibility), the public-transport service deployed in rail and bus modes, and the "fit" between the public-transport system and its metropolitan area. Public-transport service deployment in terms of vehicle-revenue miles (VRM)—the miles that vehicles travel while in revenue service—by transit mode is preferred here over public-transport investment as a metric of service, since money spent on public transport is complicated as a metric of service by the lumpiness of capital investment and the variability of costs between regions.

By making accessibility the measured outcome, we aim to provide actionable guidance for public-transport planners and delineate policy-relevant variables. Planners may choose to deploy more or less public-transport service across various technologies, and they may deploy it in a more concentrated or dispersed fashion. However, they do not directly control metropolitan size or even urban form (though land use may respond to their investments in the longer term). Policy variables here include the extent of public-transport deployment across the various modes and the fit between a public-transport deployment and its metropolitan region. The fit criterion, inspired by Robert Cervero's argument that public transport succeeds when it is designed to match the region it serves, refers to the concentration or spread of public-transport accessibility in comparison to the concentration or spread of metropolitan population.[23]

Job Accessibility by Public Transport for Metropolitan Regions

This analysis evaluates the contribution of rail and bus public-transport modes to job accessibility in forty-six of the largest metropolitan areas in the United States. Rail, as defined in this study, corresponds to what the National Transit Database (NTD) categorizes as "fixed guideway" and bus corresponds to the category of "non-fixed guideway."[24] BRT service is difficult to isolate as a separate mode and appears in this study in both fixed and non-fixed-guideway categories but is a tiny fraction of each. Service identified by its operator as BRT is present in eleven of the forty-six metropolitan areas studied and is split between the rail and bus categories depending on its right-of-way.[25]

The components of public-transport accessibility to jobs are explored here through multiple regression models with work accessibility by public transport as the outcome of interest. Data for these analyses come from the University of Minnesota Accessibility Observatory, the National Transit Database, the National Historical Geographic Information System, and LEHD Origin-Destination Employment Statistics.[26] The dependent variable for these analyses is a population-weighted metropolitan job accessibility score.[27]

Public-transport accessibility is modeled on the basis of independent variables that include the total population in a metropolitan area (as a control variable), VRM per capita for rail and bus modes, residential density (measured as a share of the regional population living in densities higher than sixty persons per gross acre), a Fit Index that assesses the degree to which the regional land-use pattern corresponds to the public-transport accessibility provided (as a correlation between block-level thirty-minute public-transport accessibility and block-level gross population density), and a Gini Index that measures the evenness of the distribution of public-transport accessibility across the regional population.[28]

The evenness of the distribution of public-transport accessibility across the regional population, as measured by the Gini Index, can affect average accessibility in two ways, with opposing directions of effects. First, concentrating public-transport resources in the highest-density and most central territory may pay off well for residents of these districts. In principle, this outcome could be enough to raise average per-capita public-transport accessibility overall. Second, spreading transit resources broadly both increases the transit accessibility of people in more outlying locations and increases the ability of centrally located people to reach those distant locations by transit. If the second effect dominates, then the prescription of fitting public transport to its region by spreading it around to residents outside the urban core makes good sense from an accessibility perspective. Note, however, that deployment policies to increase public-transport accessibility for the average metro resident may differ from those to promote social equity or those to maximize ridership, which may be increased by focusing public-transport resources on the most transit-friendly corridors in the region (see chapter 6).

Effects of Public-Transport Deployment, Land Use, and Fit on Accessibility

Regression models were estimated to assess the relative importance of several independent variables on the dependent variable of public-transport accessibility (table 5.1). The metropolitan areas in these models include regions whose public-transport accessibility has been strongly shaped by public-transport investments from the first half of the twentieth century.[29] Thus, to represent more closely the transit-investment decisions that metropolitan areas face in the current era, models were also constructed that exclude metropolitan areas with legacy heavy-rail systems, defined as heavy-rail systems (except for commuter rail) from before 1972 (these are shown in table 5.1 as Models B and D). This excluded from the analysis metropolitan New York, Boston, Chicago, Philadelphia, and Cleveland.

Two regression models use average weighted public-transport job accessibility as the dependent variable (shown as A and B in table 5.1). One way to think of these is to consider the individual enjoying the average level of public-transport accessibility in the region. Since public-transport accessibility tends to decrease from the metropolitan center to the periphery, this person is likely a suburban dweller somewhere between the metropolitan edge and the municipal boundary of the central city. These two models indicate a strong relationship between public-transport accessibility and the service provided by rail and bus modes. Unsurprisingly, public-transport accessibility depends on public-transport deployment; in combination, the two public-transport service variables are the greatest explanatory factors in these models, followed by residential density variable and land-use/transit-service fit.

Rail has a stronger influence on public-transport accessibility than bus. Rail modes contribute between 49 percent and 96 percent more than bus modes per vehicle-mile of service to the average job accessibility by public transport in a region. The difference between the share of accessibility explained by rail and bus

Table 5.1 Influence of public-transport service and land use on public-transport accessibility

DEPENDENT VARIABLE	TIME-WEIGHTED JOB ACCESSIBILITY VIA TRANSIT		85TH-PERCENTILE 30-MINUTE JOB ACCESSIBILITY VIA TRANSIT	
MODEL	A	B	C	D
ANALYSIS SET	ALL METROS EXCEPT NEW YORK	ALL METROS EXCEPT LEGACY RAIL METROS	ALL METROS EXCEPT NEW YORK	ALL METROS EXCEPT LEGACY RAIL METROS
Rail VRM per	0.39 ***	0.35 ***	0.18 *	0.19 *
capita	(.0425; 5.58)	(.0465; 4.57)	(.0698; 1.86)	(.0710; 2.03)
Bus VRM per	0.20 ***	0.23 ***	0.14 **	0.09
capita	(.0210; 4.39)	(.0221; 4.19)	(.0345; 2.21)	(.0337; 1.36)
Percent in high	0.37 ***	0.40 ***	0.70 ***	0.69 **
density	(.1342; 4.40)	(.1631; 4.41)	(.2201; 6.13)	(.2491; 6.19)
Population	0.20 ***	0.18 **	–0.04	–0.02
	(.0357; 3.55)	(.0399; 2.73)	(.0585; –0.55)	(.0609; –0.22)
Gini Index	–0.04	–0.05	–0.14 **	–0.12 **
	(.0176; –1.15)	(.0180; –1.05)	(.0288; –2.68)	(.0274; –2.24)
Fit Index	0.09 **	0.09 *	0.16 **	0.21 ***
	(.0207; 2.13)	(.0214; 1.85)	(.0340; 2.69)	(.0327; 3.37)
Intercept	(.0219; –0.33)	(.0271; 0.48)	(.0359; 2.25)	(.0415; 3.44)
Adj. R^2	0.9383	0.9224	0.8826	0.8854
N	45	41	45	41

Notes: Values are standardized beta-weight coefficients. Values in parentheses are standardized standard errors and t-statistics. Statistical significance: *p < 0.1; **p < 0.05; ***p < 0.01.

systems would be a function of both the speed differences between the two technologies and their association with land use. For example, if rail is associated with greater job-housing proximity or clustering of land uses, its accessibility impact would incorporate this. This land-use effect may explain in part why the bus-rail gap is larger in the rail-legacy metropolitan areas than in the non-rail-legacy regions. Having grown up around their rail systems, the rail-legacy metropolitan areas have had many more decades for their physical development patterns to respond to rail accessibility than the regions without historic rail investment. To the extent that rail deployment is positively associated with higher spending on buses, the contribution of rail modes to job accessibility may be even larger than what is found here.[30]

The correspondence between land use and public-transport service (as shown by the Fit Index) also contributes to the benefits provided by public-transport service and residential density. This suggests that there are regional accessibility benefits to coordinating land-use patterns and public-transport investments through efforts such as transit-oriented development and in general permitting denser development in areas of high public-transport accessibility. A broad distribution of public-transport service, as indicated by a lower Gini Index, is weakly associated with higher average public-transport accessibility, an expected result (though not statistically significant), since a higher Gini Index represents a less even distribution of public-transport services in the region. This suggests that as public-transport service coverage is spread more evenly across the population, average regional job accessibility increases.

The discussion of the two models (A and B in table 5.1) pertains to average job accessibility via public transport as the relevant outcome. Seeking to boost average accessibility is, however, only one possible transit-accessibility outcome. It may be, by contrast, that high-quality public transport at the upper end of the accessibility distribution is what matters. For example, planners and policy makers may be motivated to focus on the high end of the transit-accessibility distribution to promote the possibility of car-free lifestyles, or to improve conditions for low-income residents at the center of a region. In any case, a shift of focus to the high end of the transit-accessibility distribution implies greater attention to the center of the metropolitan area than to the suburbs.

Two additional models incorporate this idea by shifting the dependent variable to the 85th percentile of accessibility (shown as C and D in table 5.1).[31] When the focus is on this high end of accessibility in the region, rather than the mean, the explanatory power of both rail and bus service is relatively less influential, while the relative importance of residential density and land-use/public-transport fit is greater. Rail and bus are nearly equivalent in the model including legacy metropolitan areas, perhaps as a result of the relatively central location of the 85th-percentile individual in dense-core regions. By contrast, rail is much more important than bus in nonlegacy regions for improving 85th-percentile accessibility.

The geographic spread of public transport, as measured by the Gini Index, plays an important role in explaining this higher end of public-transport accessibility. In metropolitan areas where public-transport accessibility is distributed more broadly across the population, 85th-percentile public-transport accessibility increases. Whereas a broad distribution of public-transport accessibility may boost the accessibility of both the average and the 85th-percentile resident, the increase is substantially greater for the high end of accessibility.[32] Public-transport accessibility tends to be very spatially concentrated, and a broader spread, in many cases, means a particular boost for residents around the 85th percentile. In addition, a broader geographic spread implies growing public-transport accessibility overall because of the ability of people with transit-system access to reach more destinations. Given their relatively central location, the 85th-percentile residents are well-positioned to benefit from a public-transport system with expanded geographic reach.

The explanatory power of residential density increases with the shift from average-accessibility to 85th-percentile-accessibility models. High densities increase public-transport accessibility by improving proximity to destinations as well as by making it easier to serve the urban population efficiently with public-transport service. By contrast, the population variable becomes less influential in the 85th-percentile models. This may be a product of the definition of public-transport accessibility based on a thirty-minute commute time, on which these models rely. Public-transport travel of thirty minutes from central locations would still be within the metropolitan core for most large regions in the United States; hence the size of the metro is less important than the extent of public-transport service provided. At the same time, proximity is more important for a thirty-minute threshold, and that is presumably captured via the residential density variable.

Land-use/transit-service fit increases in importance moving from the average to the 85th percentile, and it has as large an influence as either rail or bus service. This highlights the importance of integrating transportation and land-use policy to improve public-transport accessibility for urban residents.

From Accessibility Measurement to Public-Transport Resource Allocation

While debates over public-transport policy are frequently framed in bus-versus-rail terms, one point emerges from our analysis: public-transport accessibility depends on investment in either technology. Metropolitan regions such as Kansas City, Phoenix, and Detroit that fail to deploy much public-transport service in either their bus or rail variants will rank poorly in public-transport-accessibility terms. In this sense, the threat to public-transport accessibility is much less the danger of making the wrong choice of public-transport technology than the unwillingness of electorates (or their representatives) to spend much money on public transport in general. And the political orientation of the electorate is among the strongest predictors of public-transport spending per capita.[33]

In bus-versus-rail terms, all else being equal, a mile of rail service is associated with an average accessibility benefit that is between 49 percent and 96 percent higher than that produced by a mile of bus service. While this finding may be interpreted as favorable to rail deployment, it is hardly the end of the story. The decision to invest in a certain public-transport technology depends not only on the benefits it is expected to generate but also on the costs associated with its construction and operation.

From a narrow cost-benefit point of view, if rail is to be preferred over bus, rail costs must not be more than 96 percent higher than bus costs. When taking both capital and operating costs into account, this condition does not hold, since the capital costs of rail are much higher than buses in mixed traffic. The U.S. Government Accounting Office estimated per-mile LRT capital costs to be fifty times those of buses on arterials.[34] Buses in mixed traffic share the cost of the bulk of the infrastructure with cars and trucks; even with the fair share of these costs attributed to the buses, the capital costs cannot be close to the same range as those associated with rail transit. Moreover, capital-cost differences are not likely to be compensated for by operating-cost savings. One meta-analysis put LRT at three times the operating cost of bus per vehicle revenue mile (though because of high occupancies, LRT showed slightly lower operating costs per passenger mile).[35] With the rail-bus cost gap much greater than the accessibility differences, the analysis presented here would appear to support greater reliance on buses in providing regional public-transport accessibility on a cost-benefit basis.

The criterion of accessibility generated per money spent is probably too narrow a guide for transit-service planning, however. First, a rail vehicle-revenue mile transports more passengers than a bus vehicle-revenue mile; in the thirty-nine metropolitan areas with rail included in this study, median vehicle occupancy in fixed-guideway modes (22.7) was about triple that of vehicle occupancy in non-fixed-guideway modes (7.5).[36] Thus, higher occupancies partly compensate for the higher costs of rail. Buses in mixed traffic generally travel at slower speeds than cars along the same route; for this reason, they rarely provide a motive for people with access to autos to choose the public-transport option. Corridors with high bus occupancy can increase the person throughput of their facilities through designation of a dedicated guideway; a BRT lane has a carrying capacity of 5,000–45,000 passengers per hour.[37] Given a high enough volume of public-transport passengers, this compares very favorably with the maximum throughput for mixed traffic: 1,900 vehicles per lane per hour of green for signalized intersections.[38] Public-transport planners who seek to provide high-passenger-volume corridors and to increase public transport's attractive capacity, as compared to the private car, may advocate for increased reliance on fixed-guideway systems, including BRT and rail.

Moreover, the net effect of a decision to opt for expensive rail transit on public-transport accessibility would need to be assessed in light of the political processes by which revenue for public-transport investments is raised. If spending on fixed-guideway transit reduces spending on bus transit, the decision to allocate resources to LRT (or even BRT) could reduce accessibility. Those resources, if spent on adding ordinary bus services, might augment public-transport accessibility more. But in practice, areas that spend more money on rail end up spending more money on bus.[39] Allocation between bus and rail spending thus does not amount to a zero-sum game; it may be that rail spending supports the creation of both transit-friendly areas and transit-friendly constituencies. In this environment, the substantial accessibility advantage offered by fixed-guideway public-transport service over bus service may help explain rail's continued appeal.

Yet resource-allocation decisions in either bus or rail remain only part of the transit-accessibility story. While public-transport investment is central to public-transport accessibility, the results presented here suggest that accessibility can in many cases be degraded through poor land-use planning. Limiting residential concentrations in areas that are well-served by existing public-transport service—a common move driven by mobilized local constituencies—can reduce public-transport accessibility to jobs for the average metropolitan resident as well as for the central-city resident.[40] In fact, for centrally located residents, the variation in the public-transport accessibility provided is explained to a greater degree by residential density and land-use/transit fit than by the extent of public-transport service. In an era of resource constraints, land-use regulatory reform to relax density restrictions in areas of high public-transport accessibility may be the most promising approach to boosting public-transport accessibility.

Making the Shift to Accessibility-Based Planning for Public Transport

Public transport differs from other modes when it comes to accessibility. Like its counterpart modes, public transport provides mobility that can contribute to accessibility. But in ways distinct from other modes, public transport can exert a stronger influence on proximity to improve accessibility. Spatial and temporal characteristics of the public-transport mode impose special challenges on planners and policy makers who aim to make the shift to accessibility-based evaluation in public transport. Although choosing to take a simple approach to address these technical challenges may sacrifice some accuracy in estimating, the payoff can nevertheless be substantial if accessibility measurements lead to policy insights not discoverable through mobility-based evaluation alone.

The accessibility shift requires that accessibility become the core rubric by which public transport is planned and evaluated. The multiregion empirical study presented here provides an example of the different perspective on public-transport decisions afforded by the accessibility shift. When increasing public-transport accessibility is the goal, large regions have good reason to increase transit deployment, though the choice between rail and bus technologies is a complex one based on the interplay among accessibility impacts and local financial and political trade-offs. Coordinating land-use development along with public-transport investments—through transit-oriented development or easing of low-density zoning restrictions, for example—leads to regional accessibility benefits. Distributing public-transport service broadly and evenly throughout a region brings benefits even to residents at the center of the region by providing them with an expanded geographic reach.

Single-region evaluations of transit accessibility can be brought even closer to decision points than the multiregion evaluation. A public-transport operator with good accessibility models can use them to test the impact of potential service changes, such as relocations of bus routes or increases in frequency, on the overall level and equity of distribution of transit accessibility. Accessibility-based analyses like these can be used to shape public debate and inform decision makers about which combination of transport and land-use arrangements best helps a region's public-transport riders reach their destinations.

6

ACCESSIBILITY IN SOCIAL-EQUITY EVALUATION

A long-standing challenge for transportation planners is the search for a fair distribution of the costs and benefits of a transportation system. Gauging transportation equity involves a range of dimensions, including exposure to negative consequences like noise and air pollution, distributions of monetary and other costs, tax and subsidy incidence, and disparities in safety outcomes.[1] These dimensions are costs borne disproportionately across populations, and they are apart from transportation service itself. In conducting equity analyses, planners typically rely on mobility-based tools, which address the distribution of costs alone. By contrast, the accessibility shift can support equity analysis based on the distribution of benefits from transportation service in the form of the ability to reach destinations. The accessibility shift, in and of itself, neither produces equitable outcomes nor inherently defines what would count as transportation fairness. But it offers planners a crucial capability, based on the simultaneous consideration of travel mode and geographic location, to evaluate how the benefits from transportation service are distributed among people; standard mobility-based metrics cannot do this.

In this chapter, we argue that because the concept of accessibility enables comparison of transportation's benefits among social groups, it is essential to proper evaluation of the social-equity impacts of transportation decisions. Unlike mobility metrics, which tend to focus on the performance of infrastructure, accessibility metrics are readily analyzed with regard to specific groups of people, such as low-income and racial-minority groups. One group that deserves special attention in equity analysis is people without access to an automobile, because the travel modes available to individuals are the single most decisive factor in determining whether they can reach destinations. But because accessibility is inherently multidimensional and more complex than standard mobility-based metrics, several methodological considerations are essential for meaningful analysis; considering differences in mode alone is not sufficient. The influence of mode on accessibility tends to vary systematically in metropolitan space, and accessibility analysis must consider the effect of mode and location simultaneously. We will first explain the mobility nature of current equity evaluation, and then propose and demonstrate the use of accessibility-based equity-evaluation tools.

Problems with Mobility-Based Equity Evaluation

When planners rely on mobility-based measures of transportation outcomes they face three problems for evaluating the distributional impacts of transportation decisions. First, mobility-based measures such as levels of congestion are attributes of transportation links and not of people. Consequently, all users of a roadway are treated as if they have the same level of mobility, a fact that hides variations in people's ability to reach their destinations. Moreover, such measures ignore households without cars, the people who are most disadvantaged by transportation conditions today.

Second, achieving congestion relief through added highway capacity may undermine accessibility by causing a spread of origins and destinations.[2] Travel to increasingly remote shopping or work destinations might be accomplished at higher speeds, but the geographic spread of these destinations demands yet more travel to achieve the same level of interaction. Thus, transportation policy may disproportionately harm racial minorities and low-income people who live near the urban core and have few resources to adapt to spreading land-use patterns.[3]

Finally, mobility-based metrics fail to capture the benefits of a transportation system, even when they are used to gauge the disadvantage that some social groups experience through their travel behavior. Equity evaluations commonly compare groups on mobility metrics such as kilometers traveled per day, trips per day, and minutes traveled per day.[4] The problem with such metrics is that it is not clear whether greater movement constitutes an advantage or disadvantage. All else being equal, one would prefer a shorter travel time to a longer one. However, a preference for shorter travel time does not mean that those with longer travel times are somehow disadvantaged. In fact, social groups that are widely considered to be disadvantaged in transportation, such as women and poor people, demonstrate much shorter travel times than the general population mainly because they have fewer choices.[5] By contrast, most of the middle- and upper-income classes of the United States can choose to trade off longer commutes in exchange for the amenities that come with living in the suburbs, and the longer travel times or distances they experience would not properly be considered a transportation disadvantage.

Equity Analysis in Professional Practice

Supporting equity in transportation is more than a moral or professional obligation; it is also mandated by a series of laws and regulations in countries, including the United States, that require ongoing and active monitoring by the public agencies charged with transportation planning. Yet, despite these mandates, public agencies charged with providing transportation infrastructure and services struggle with the task of evaluating whether their decisions comply with equity objectives.[6] The conceptual difficulty of evaluating equity has resulted principally from reliance on mobility-based metrics; many of these challenges can be overcome with a shift to an accessibility framework.

The Challenge of Meeting Equity Mandates

In the United Kingdom, a national goal of reducing social exclusion became a top public policy objective in the 1990s and prompted government officials to advance efforts to address inequality in the transport sector.[7] Recognizing the inequitable distribution of opportunities to participate fully in societal activities, policy makers and researchers promoted changes in the provision of transport as an important factor in enabling people to undertake a wider range of activities in daily life. The UK Social Exclusion Unit published a widely recognized report in 2003—which influenced practitioners and scholars in several countries, most notably the Netherlands and Australia—about the importance of addressing the role of transport through a framework of overcoming social exclusion.[8] In 2006 transport authorities in England at the local level of governance were charged by statutory requirement to identify solutions to the problem of individuals and communities experiencing low accessibility to essential destinations.[9] Even though a change in national government and economic austerity measures limited the capacity of local authorities to implement solutions, Karen Lucas concludes that "the concept of transport-related social exclusion now has significant resonance with the transport research and policy community internationally."[10]

In the United States, state and local agencies are legally required to avoid discrimination in their plans and programs. This mandate has its roots in a series of laws including Title VI of the Civil Rights Act of 1964, the National Environmental Policy Act of 1969 (NEPA), and several Federal-Aid Highway Acts of the 1970s.[11] Partly in response to the growing environmental-justice movement and the need for public agencies to play a more assertive role, the Clinton administration in 1994 substantially elevated attention to social equity when it issued Executive Order 12898 (Federal Actions to Address Environmental Justice in Minority Populations and Low-Income Populations). The executive order, which remains in effect as of this writing, directed all federal agencies to develop a strategy that "identifies and addresses disproportionately high and adverse human health or environmental effects of its programs, policies, and activities on minority populations and low-income populations."

Despite the growing awareness of inequities in the transport sector and the demands on governments to address them, public agencies often find these requirements challenging to implement due to a mix of technical and political considerations. In the United States, guidelines are vague and the requirements rarely enforced, which commonly results in highly incomplete analyses.[12] Frequently, legislation and regulations fail to specify evaluation techniques to ensure that the principles outlined in the various environmental-justice goals are met, and planners and decision makers have expressed the need for such technical tools.[13] Similarly, prior to the United Kingdom's mandate to carry out accessibility planning to address social exclusion, local authorities gave equity analysis little priority or relied on overly technical analyses that failed to yield policy change.[14]

Mobility-Based Equity Analysis in Professional Practice of the United States

Equity analysis in professional practice remains rooted in mobility thinking in the United States. The most specific guidance on carrying out equity analysis has come from the Federal Transit Administration (FTA), providing instructions for all recipients of financial assistance from the FTA, including state departments of transportation, metropolitan planning organizations (MPOs), and public-transit agencies.[15] Recipients of federal funding must certify their compliance with civil rights law. They must have a planning process, according to FTA guidelines, that "identifies the needs of low-income and minority populations" as well as an "analytical process that identifies the benefits and burdens of metropolitan transportation system investments for different socioeconomic groups, identifying imbalances and responding to the analyses produced."[16]

In the case of public transport, agencies are required to maintain system-wide service standards and to perform an equity analysis of proposed changes in service or fares, including alterations to routes or schedules and the elimination of bus lines or stops. If an agency finds disparities, it is required to take corrective action. The recommended approach for demonstrating compliance is to create maps showing the proximity of minority and low-income population groups to bus and rail stations. Yet proximity to a transit stop is merely an indicator of access to the public-transport system and not a measure of accessibility to destinations (see chapter 5). Under the recommended approach, service levels are to be evaluated on a range of metrics including vehicle loads (e.g., passengers per vehicle), headways (the frequency of service as the time interval between vehicles arriving at a stop), on-time performance, and availability of amenities (e.g., benches, shelters, trash receptacles). Service metrics like these gauge important dimensions for agencies to consider, but they are ultimately mobility based because they can only assess disparities in how well vehicles are moving or how people are experiencing the transportation system. They do not address disparities in how well public transport is meeting its ultimate purpose of connecting riders with their destinations.

In the case of regional planning agencies, the most common type of equity analysis is conducted by MPOs to certify that their regional plans comply with civil rights law. The FTA's widely used approach first identifies geographic concentrations of population groups including, at a minimum, racial-minority and low-income residents of the region.[17] Typically, two groups are defined so that their metrics can be compared. One group is often termed the "EJ (Environmental Justice) population" (e.g., a minority or low-income population, which is sometimes called a "protected population" or a "target population"). The EJ population is then compared to a control population (e.g., a nonminority or non-low-income population).[18]

This type of analysis defines a set of metrics, typically mobility based, with which to evaluate the benefits and burdens of the regional transportation plan. In some cases, MPOs use mobility metrics alone, without any acknowledgment of accessibility. An example is from the Mid-America Regional Council, which serves the Kansas City, Missouri, metropolitan area.[19] Examples of mobility-based metrics used in the analysis include the number of trips taken by auto and transit, walk time to a transit stop, and hours of transit service provided. After comparing these and other metrics between the population groups, the report finds that the proposed regional plan will have no disproportionately high or adverse impacts on the protected populations.

Environmental-justice analyses have started to use some accessibility metrics, typically as a complement to mobility-based analyses.[20] Several MPOs exemplify the use of accessibility metrics in equity analysis, including the Boston Region Metropolitan Planning Organization, the Metropolitan Transportation Commission of the San Francisco Bay Area, the Mid-Ohio Regional Planning Commission of Columbus, and the Southern California Association of Governments.[21] Yet the accessibility metrics appearing in these MPOs' equity analyses remain merely supplements to a mix of mobility-based metrics.[22]

The Mid-Ohio Regional Planning Commission of Columbus, Ohio, provides an example of an extensive and thorough environmental-justice analysis that acknowledges the importance of accessibility but also relies heavily on mobility-based metrics. To illustrate how a mobility-based metric is assessed in the plan, the analysis includes a chart comparing a 2035 "no build" scenario to a 2035 scenario of plan implementation, and it shows that the share of congested vehicle-miles of travel is expected to be higher for the target disadvantaged population relative to the nontarget population in both scenarios.[23] Such a finding is not surprising, because the target population is located at the core of the region, precisely where traffic congestion is typically highest. But the finding that one group would experience more congestion on average than the other group offers little insight into the question of which group experiences disadvantage. It may be that higher levels of congestion are offset by higher levels of accessibility. But the relevant equity question is multidimensional and more complex than even measurement of the accessibility of different zones. For example, centrally located territory may be more accessible than the metropolitan periphery by either auto or public transport, but its residents may experience low accessibility because of their low car-ownership rates. Similarly, the distribution of public-transport accessibility may match the location of transit-dependent people well, or there may be systematic disjunctures between the two.

Advancing Equity Analysis with an Accessibility Perspective

While mobility-based metrics leave uncertainty about whether social groups experience disadvantage relative to others, accessibility metrics provide a logical starting point for equity analysis.[24] While arguing for the normative dimension of accessibility, Martin Wachs and Gordon Kumagai advocated for the relevance of such analysis to the equity-based assessment of conditions among social groups.[25] In fact, some scholars have argued that accessibility is the appropriate good to be considered in questions of just distributions of transportation benefits.[26]

Scholars using accessibility in social-equity analysis have made important advances in at least two respects.[27] First, they take a normative stance on distributive justice based on the distinction between "horizontal" and "vertical" equity.[28] These studies start from the observation that horizontal equity, under which members of any one group, such as poor people or car owners, are expected to experience identical benefits, is all but impossible because of the spatial nature of accessibility. Attributes of places within metropolitan space, such as residential density, demographics, destinations, and proximity to the center, vary so substantially that any location-based benefits such as accessibility are unavoidably unequal. Designing transportation services in a way that achieves identical levels of accessibility for everyone is insuperably difficult in this respect. But delivering identical levels of accessibility for everyone through transportation may not be desirable even if it were possible to achieve. The concept of vertical equity expresses a normative approach to deliberately distributing benefits unequally among population groups according to principles such as accommodating preferences, considering who has the ability to bear costs, or meeting needs. These studies take the view that transportation benefits ought to accrue to some groups more than to others in order to meet needs or address preexisting disadvantages.[29] Clearly, how public officials define social groups and decide which principles to follow are not merely technical decisions but value-laden ones.

The second main advance in studies of accessibility and equity is an understanding of the need to establish principles to guide analysis in practice. Even if transportation professionals have started to accept the concept of accessibility as a viable equity indicator, there remain wide variations in how it might be measured and evaluated.[30] For example, no standards govern the equity analyses that are mandated by federal regulations in the United States.[31] Karel Martens proposed principles to "assist decision-makers in the selection of transport projects that maximize average accessibility levels, while ensuring that the accessibility gaps between population groups remain within an acceptable range."[32] These principles properly identify the goal of increasing accessibility as the main objective of transportation planning, and they aim to make sure that disadvantaged populations are not harmed by constraining the range between the best-off and worst-off in terms of accessibility.

One manifestation of these principles is in the call for minimizing the gap between auto and public-transport accessibility.[33] The goal of minimizing the difference between transit and auto accessibility stems from the car's status as the yardstick for a base level of transportation convenience in an auto-oriented society:[34] Karel Martens, Aaron Golub, and Glenn Robinson contend that using the transit-auto accessibility gap should be a key indicator: "Ideally, transport investment programs should guarantee that: (1) the gap between the areas or neighborhoods with the lowest and the highest level of access should remain within a predefined range . . . ; (2) the gap between car-owning and car-less households residing in the same area or neighborhood should remain within a predefined range . . . ; while (3) aiming to achieve the highest possible average access level across all neighborhoods and mode-related groups."[35]

Minimizing the gap between public transport access and auto access is advocated out of concern for several disadvantaged groups: those who cannot afford to own a car, those who cannot drive for physical or other reasons, and those who are too old or too young to drive but still want to participate in activities. Proposing to minimize the gap between auto and transit accessibility is sensible to the extent that it aims to address the needs of the most disadvantaged, and because it appropriately recognizes that relative deprivation is an important element in social justice. People may be disadvantaged not just in an absolute sense, but also in comparison to a societal norm.

Yet the principles of providing preferential treatment for certain groups like transit-dependent people and seeking to minimize the gap between auto and public-transport accessibility in any given area or neighborhood, are likely to be at odds with one another. This is because of two intersecting phenomena: public-transport users are highly unequally distributed from area to area, and the cost of providing transit access varies markedly from central to peripheral areas. For example, allocating a given budget so as to minimize transit-auto accessibility gaps for all areas would lead to an entirely different spending pattern from an allocation oriented to maximizing the accessibility of transit-dependent people. The latter principle would demand focusing investment in areas where individuals without access to cars reside. In many instances, those are also the higher-density, central locations that are served most economically by public transportation in any case. Thus, promoting the accessibility of people without cars will often require a targeting of public-transport resources that leaves some areas with large gaps between transit and auto accessibility. A focus on the gap between transit and auto accessibility across the region as a fairness indicator amounts to a definition of equity based on territory. By contrast, we will offer approaches that aim to focus on accessibility through explicit accounting for the residential locations of people with and without access to cars.

Accessibility Analysis by Mode, Location, and People

Costs and benefits of transportation infrastructure and services vary across metropolitan space, individuals, and households. In the case of U.S. metropolitan regions, nearly all people without cars experience a substantial accessibility disadvantage relative to their car-owning counterparts. Planners seeking to target transit resources where needs are greatest must simultaneously consider the effects of mode, location, and people when assessing the distribution of accessibility benefits.

How Accessibility Varies by Mode and Location

The magnitude of the gap between auto and public-transport accessibility is striking in nearly every metropolitan region of the United States, and the disparity between the modes varies systematically within metropolitan space. For example, even Washington, DC, a metropolitan area that has an unusually strong commitment to public transport, demonstrates a substantial gap between travel modes. Figure 6.1 compares the regional accessibility of Washington, DC, to Detroit by auto and transit, and highlights several observations about the importance of analyzing accessibility by mode and location simultaneously. First, both auto and public-transport accessibility show generally similar patterns: the values are highest near the central business district (CBD) and then taper off to become lowest at the periphery of the region. Second, transit accessibility is low even in the CBD, where public-transport service is most prevalent, in comparison to auto accessibility at the same location. Third, transit accessibility is low throughout the majority of metropolitan space. In Washington, DC, public-transport accessibility drops steeply within about eight kilometers of the CBD and becomes close to zero after about twenty kilometers. The difference between the best location for public transport (0 km) and the worst (100 km) is small and represents just a fraction of the full range between the best and worst locations for auto.

While people with access to cars have a significant accessibility advantage virtually everywhere in U.S. metropolitan regions, the extent of that advantage varies with the type of region and the public transport investment it has made. For example, in contrast to transit-rich Washington, DC, the Detroit metropolitan region is well known for poor public transport service (figure 6.1).[36] Detroit exhibits patterns of accessibility that are more similar to those of other U.S. metropolitan regions than to those of Washington, DC. All the observations noted for Washington, DC, hold for Detroit, but the difference between auto and transit accessibility is more extreme in Detroit. Furthermore, the Detroit case makes clear another contrast between auto and public-transport accessibility that generally holds for most U.S. regions. Accessibility by auto is substantially better than accessibility by public transport nearly everywhere. Car drivers as distant as seventy-five kilometers from the CBD enjoy higher levels of auto accessibility than transit riders located at the CBD where transit accessibility is highest.

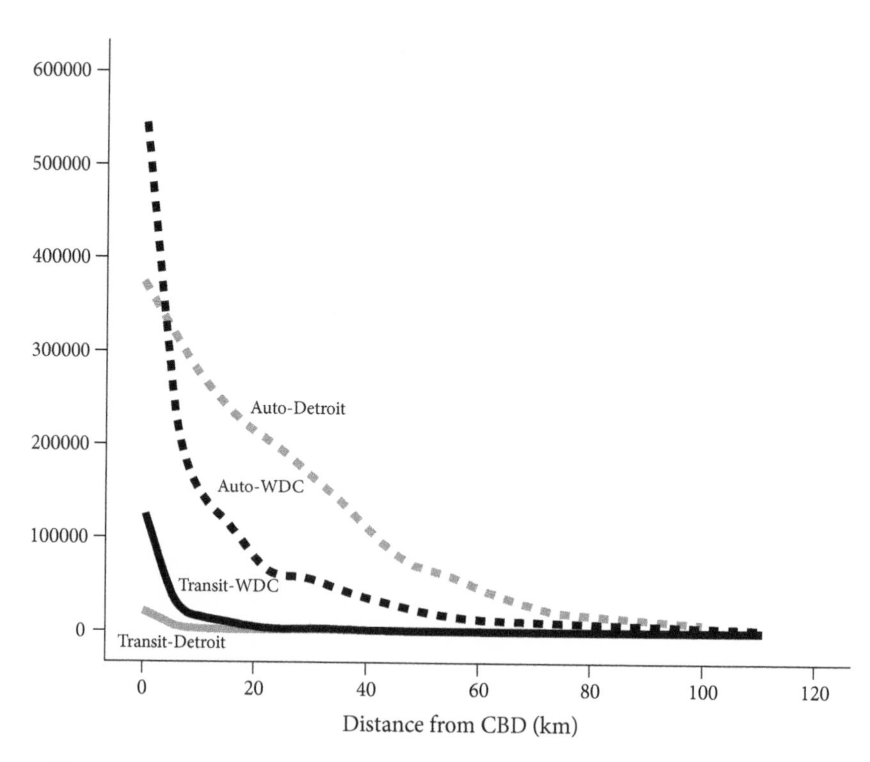

FIGURE 6.1. Comparing automobile with transit accessibility to work, by distance from the central business district (km), Washington, DC, and Detroit MPO regions, 2005

These charts illustrate what Evelyn Blumenberg and Michael Manville refer to as the "modal mismatch, a drastic divergence in the relative advantage between those who have access to automobiles and those who do not."[37] The severe difference between auto and public-transport accessibility in U.S. metropolitan areas is the defining feature of transportation disadvantage. Any policy that aims to address social equity in transportation must confront the conditions of transit-dependent people. Policy interventions such as strengthening public transport, subsidizing auto ownership for poor people, and fostering alternatives to car ownership through car-sharing or bike-sharing programs are examples of steps that would elevate the accessibility of transit-dependent people.[38]

Effectively promoting social equity in transportation requires tools to help decision-makers choose where to invest the next increment of transportation funds, and policies to guide investments without explicit attention to where and at whom the investments are aimed can foster perverse outcomes. For example, in Detroit, planners seeking to minimize the gap between auto and public-transport accessibility in each neighborhood would be obligated to invest in transit at locations that are some thirty kilometers from the CBD (figure 6.1). While spreading transit service broadly can increase accessibility at the metropolitan center (see chapter 5), the effect would depend on the destinations served by suburban extensions. Where these serve mostly residential zones, they would only have a slight effect on the accessibility of central-city residents; by contrast, where they serve concentrations of work or nonwork destinations, they can be a significant accessibility boost. For this reason, tools are needed to ensure that resource allocation benefits the people who rely on public transport, including those who live at the metropolitan center.

Transit-resource allocations that appear to fail at that goal have led to concerns that investments aimed at "choice" riders in the suburbs are harming "transit-dependent" riders at the urban core.[39] Prioritization of public transport away from the urban core in U.S. cities in favor of suburban jurisdictions has triggered a wide array of legal challenges based on the claim that such funding favors predominantly white and relatively affluent populations over racial minorities and the poor.[40] If steps to consider social equity are to address the severe differences between auto and transit accessibility, then guiding principles must account for transit-dependent people and the locations where they live.

Analyzing Mode and Location by Social Groups

Without more information, figure 6.1, which pertains to mode and location simultaneously, reveals little about people, notably people who depend on public transport. Transit-dependent people are not evenly distributed in metropolitan space but, in the case of the United States, are typically concentrated near the metropolitan core, precisely where public-transport accessibility tends to be high.[41]

Regardless of the level of public-transport accessibility provided in a region, carless people can increase their accessibility by locating in zones of higher transit accessibility. Ideally, the modal disadvantage of the transit-dependent population would be at least partly compensated by their locational advantage if they resided in zones of exceptionally high public-transport accessibility. By contrast, transit-dependent people who live in low-transit-accessibility zones are disadvantaged by both mode and location.

For this reason, one indicator of transportation equity is the extent to which transit-dependent people live in transit-rich areas. A spatial match in this dimension can result from focusing public-transport investments in areas of public-transport dependence, from the location of transit-dependent individuals in transit-rich areas, from people's choice to forego car ownership in transit-rich zones, or from any combination of the three. Land-use policy can be important in this regard. A region that allows for increases in supplies of affordable housing in transit-rich areas will tend to demonstrate a better match than one that zones in a more exclusionary fashion.

To illustrate the variation among metropolitan areas in terms of the degree to which transit-dependent people live in transit-rich territory, figure 6.2 charts the share of transit-dependent households in a region according to their residence by public-transport accessibility quartiles, with quartiles defined internally to each metropolitan area.[42] Regions that perform best in the match between transit-dependent households and territory of high public-transport accessibility include Kansas City, New York, and San Francisco. By contrast, places like Houston, Los Angeles, and Cleveland perform less well because a substantially smaller share of transit-dependent households resides in zones of the highest transit-accessibility quartile. An indicator like this could focus policy-making attention on the location

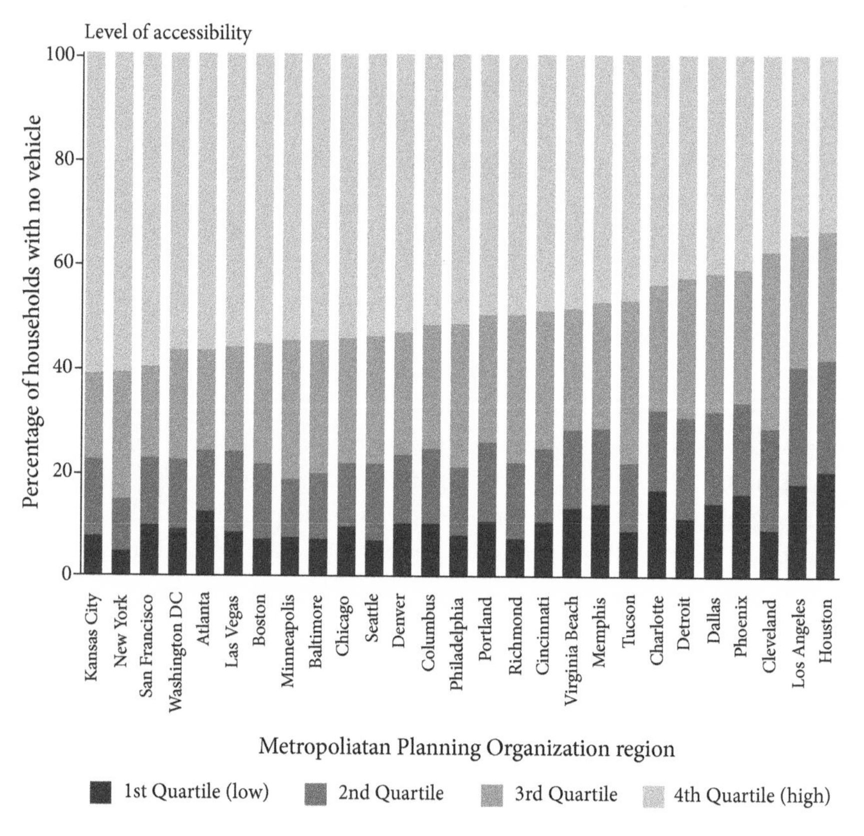

FIGURE 6.2. Share of transit-dependent households residing in four transit accessibility categories (sorted in increasing order on the highest transit-accessibility quartile)

of public-transport lines, on housing-stock constraints, or on other barriers impeding the location of transit-dependent households in transit-rich areas. One desirable attribute of this indicator is its automatic scaling to the public-transport accessibility of any given region. For example, while analysis of Kansas City's overall public-transport accessibility in comparison to New York's makes little sense, the indicator in figure 6.2 shows that the two have achieved a similar match between public-transport accessibility and transit-dependent populations.

The Mode-Location Match Ratio

The charts presented so far have shown the disparity in accessibility by mode and location alone. A quantitative indicator can be created to take the information in these charts one step further by accounting for the locations of car-owning and carless people. The "Mode-Location Match Ratio" (MLMR) is defined as the sum across all zones in a region of (a) public-transport accessibility weighted by the carless population divided by (b) auto accessibility weighted by the car-owning population. Equation 6.1 formally defines the indicator:

$$MLMR = \frac{\sum_{i}^{N}\left(a_i^t * s_i^t\right)}{\sum_{i}^{N}\left(a_i^a * s_i^a\right)} \qquad (6.1)$$

Where MLMR is the Mode-Location Match Ratio for a metropolitan region; a_i^t is the accessibility index for travel by public transport for zone i; a_i^a is the accessibility index for travel by automobile for zone i; s_i^t is the share of the region's transit-dependent households located in zone i; s_i^a is the share of the region's automobile-owning households located in zone i. For a metropolitan region with N zones, i = 1, 2, . . . , N.

This weighted ratio is highest when locations of high public-transport accessibility coincide with high shares of the region's transit-dependent population. The indicator would ordinarily range from 0 to 1, with the high values being produced by high levels of public-transport accessibility and high proportions of transit-dependent populations in the most accessible zones. Degradation of either variable would reduce the value of the indicator, and as a practical matter, in the U.S. context, the indicator would rarely exceed 0.4. The MLMR registers public-transport accessibility to the extent that transit serves the transit-dependent population. If a region chooses to improve public transport in the suburbs in a way that fails to benefit the accessibility of the transit-dependent population, the MLMR change would be slight, indicating a small impact for such a choice. By contrast, a small improvement aimed at the locations of the transit-dependent population could have a greater positive effect on the indicator.

The Mode-Location Match Ratio for a selected set of regions, in order of descending values of the MLMR, is shown in figure 6.3. Within this set, the regions with the best conditions for transit-dependent people include Washington, DC, New York, and Atlanta. By contrast, places like Cincinnati, Detroit, and Houston offer the most unfavorable conditions for carless people.

A high MLMR value can be achieved by (1) improving public-transport accessibility overall relative to auto accessibility, (2) ensuring that transit accessibility is provided where transit-dependent people live, or (3) enabling transit-dependent people to relocate to transit-rich zones by ensuring that adequate affordable housing is available there. The second and third steps are important. To gain a sense of their magnitude of influence, consider the effect of step 1 alone. In addition to the MLMR, figure 6.3 includes an unweighted ratio, a value that compares public-transport accessibility to auto accessibility by territory alone, with no weighting to account for where carless and car-owning people

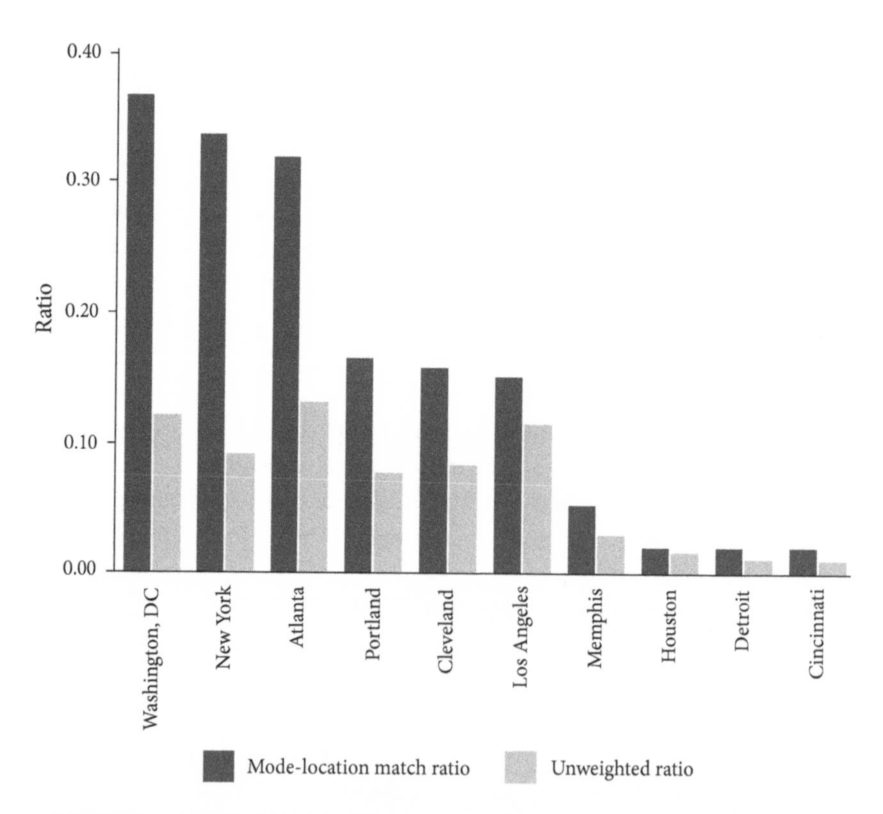

FIGURE 6.3. Mode-location match ratio, compared to unweighted ratio, selected MPO regions

live. For example, referring back to the charts reveals that the gap between auto and public-transport accessibility is much smaller in Washington, DC, than it is in Detroit (figure 6.1); thus, transit riders in Detroit are generally far more disadvantaged relative to their car-driving counterparts than are transit riders in Washington. The unweighted ratio in the chart offers a quantitative indicator to capture what the charts reveal visually: Detroit's unweighted ratio of 0.01 is substantially below Washington, DC's unweighted ratio of 0.12.

Comparing the MLMR with the unweighted ratio in figure 6.3 reveals the importance of accounting for the residential locations of transit-dependent people. First, Los Angeles is an example where transit in general appears to compete relatively well with the auto by territory alone (with an unweighted ratio of 0.11). But, consistent with figure 6.2, transit is not serving a high share of transit-dependent people in Los Angeles, as shown in the modest MLMR of 0.15 and the small difference between the MLMR and the unweighted ratio. Second, New York is an example of just the opposite; public transport in general fares much worse than the auto (unweighted ratio of 0.09) but makes up for this disadvantage by providing a high share of the best public transport to locations that house a high share of the region's transit-dependent population (MLMR of 0.34). Third, Washington, DC, appears to do well on both scores, offering strong public-transport accessibility relative to auto accessibility (unweighted ratio of 0.12), while also boosting the effectiveness of that transit by serving a high proportion of transit-dependent people (MLMR of 0.37). Finally, Detroit and Cincinnati appear to fall short on both conditions. Their public-transport accessibility is generally low relative to their auto accessibility (MLMR of 0.02 for both), and they do not direct public-transport service in a way that covers a high proportion of the transit-dependent population (unweighted ratio of about 0.01 for both).

Dependence on public transport usually constitutes an accessibility disadvantage, but dependence on public transport while living where transit accessibility is low is a double burden. The approach presented here emphasizes two main factors that contribute to the equity of accessibility among a metropolitan region's residents. First, regions where public-transport accessibility is high relative to automobile accessibility are more equitable than others. Second, a region is more equitable if a larger share of transit-dependent people is capable of living in zones where public-transport accessibility is high. This second point has critical implications for public policy aimed at advancing social equity in transportation. It underscores how land-use regulations and housing policies might contribute to improving transportation equity by relaxing restrictions on affordable-housing development in high-transit-accessibility locales and targeting the preservation and development of affordable housing in such locales.[43] And it offers clearer direction for public officials who are deciding where to invest public-transport funds. If advancing social equity is a goal, then instead of minimizing the gap between auto and transit accessibility overall, decision makers should give preferential treatment to areas of metropolitan space where large shares of transit-dependent people live when deciding where to make investments in public transport. This is all the more necessary given highly constrained public transport budgets.

Accessibility Metrics for Social-Group Comparison

While the MLMR is sensitive to locations of transit-dependent people, as a single metric, it does not convey a distribution across a population and across metropolitan territory. By contrast, analysis of distributions can highlight areas of advantage and disadvantage of population segments, such as by ethnicity or income, sometimes revealing unexpected outcomes. Equity analysis pertains to these groupings in part because public agencies in the United States have been mandated to address the needs of travelers specifically by race and income.[44]

An effective analysis of social equity requires that accessibility indicators be comparable on a common scalar across social groups. But prevailing metrics of accessibility, such as the cumulative-opportunities approach or the gravity model, present a key obstacle to making such comparisons, because they are limited to the analysis of a single mode at a time. These metrics fail to account for the differing use of travel modes among social groups, such as those defined by race and income. For example, racial minorities depend on public transport far more than whites; African Americans are about eight times as likely as whites to make a trip by bus.[45] To compare social groups, it is necessary to control for the fact that some groups depend disproportionately on the public-transport mode.

Accessibility indicators are frequently created for a spatial zone rather than for individuals. As place-based measures, they attribute the same level of accessibility to every person in a zone, regardless of personal preferences or capabilities regarding mode of travel. The typical approach to calculating accessibility indicators is to treat travel modes separately, using one indicator for travel by auto and another indicator for travel by transit. The primary reason for calculating separate indicators is that the travel-time difference between the modes impedes their meaningful combination into a composite metric.

The key to making measurable comparisons across social groups while controlling for modal differences is to conduct the analysis at the level of a household.

For our measures, each household is assigned a place-based accessibility score from the zone of residence. A household is assigned one of two accessibility scores—auto or transit—depending on the household's vehicle availability.[46] Any household without a vehicle is assumed to be dependent on public transport, and such a household experiences only transit accessibility. Conversely, any household with a vehicle available is assumed to experience auto accessibility, and every member of the household is assumed to experience auto accessibility. This assumption, which is driven primarily by data availability, suffers from certain shortcomings. How any one person among several in a household experiences accessibility is clearly more complicated than this assumption suggests. People in carless households are not necessarily dependent on transit; they may share rides with car owners or restrict their housing locations to be within walking distance of their destinations. Conversely, all people who live in a household with a car do not necessarily use that car. In a household where the number of workers exceeds the number of cars, some members may be forced to rely on transit.

Accessibility Equity and Household Income

The relationship between accessibility and income is complex. Although one might expect low income to consistently translate into an accessibility disadvantage, this is not necessarily the case. Several dimensions of the accessibility concept vary systematically by income, including the availability of a car and residential location. People with low incomes may be unable or unwilling to pay the high costs of owning a private automobile. Such people are more highly dependent on other modes of travel for accessing opportunities and tend to be disproportionately affected by policies that favor mobility for cars and contribute to the decentralization of land-use patterns. On the other hand, low-income people in the United States often live in more central parts of metropolitan areas, a location that can confer on them an accessibility advantage, particularly if they own a car.[47]

These factors combine to shape a household-level, mode-adjusted employment accessibility by income group, as exemplified in the case of Baltimore (figure 6.4). In Baltimore, about half of the households in the low-income group experience lower levels of accessibility to jobs than the other income groups. Low-income households at the low end of the accessibility range are mainly car-owning households located in zones at the periphery of the region and transit-dependent households located anywhere in the region. However, the low-income line crosses the high-income line around the median. Although low-income households are disadvantaged in the low range of accessibility, at the high end of accessibility it is just the reverse: low-income households are advantaged relative to their high-income counterparts in the high-accessibility range. Low-income households at the upper end of the accessibility range are mainly car-owning households that experience a locational advantage over their high-income counterparts because they reside near the center of the region.

Analysis of sectoral accessibility distributions as presented here can forestall misdiagnosis of problems. For example, with a significant share of the low-income population actually advantaged by their location, spatial mismatch may be less of a problem than the threat to accessibility posed by displacement from central locations stemming from reductions in housing affordability.[48] Most U.S. metropolitan regions reveal patterns that are similar to the case of Baltimore, highlighting the disparate levels of accessibility disadvantage faced by low-income households.[49]

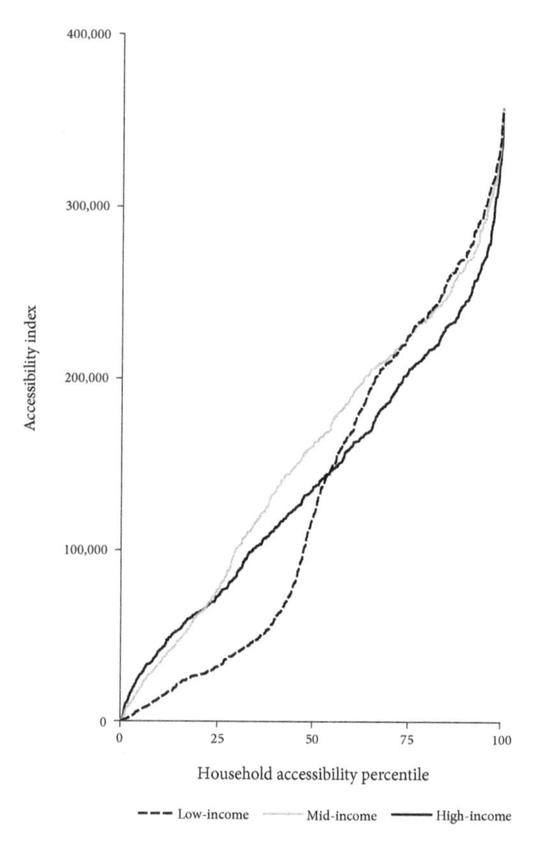

FIGURE 6.4. Mode-adjusted accessibility to work by household accessibility percentile and household income, Baltimore

Accessibility Equity and Race

The components of accessibility vary considerably by race. Racial minorities tend to own fewer cars and rely more heavily on public transport, which suggests that they experience an accessibility disadvantage.[50] Racial minorities are also not evenly spread throughout metropolitan space; African Americans in particular tend to be disproportionately located at the urban core in many U.S. metropolitan regions due in part to exclusionary zoning practices and historical racial discrimination in housing markets.[51] Residing in a central position in metropolitan space may be an advantage in terms of accessibility. Whether this locational advantage is enough to offset the mobility disadvantages of carlessness is an open question. For instance, while transit dependency is ordinarily a substantial disadvantage in accessibility, some metropolitan regions may provide transit service at a level that nearly compensates for this disadvantage. And some metropolitan regions may offer a larger share of people in need of good transit service the option of living in accessibility-rich places.

Analysis of accessibility by race can be carried out for any census-defined groupings; the following graph compares accessibility between Asian, black, and white populations in selected U.S. metropolitan areas. For example, the

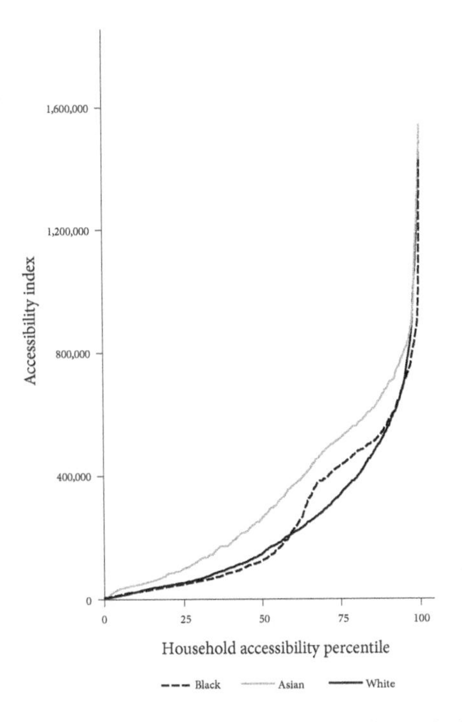

FIGURE 6.5. Mode-adjusted accessibility to work by household accessibility percentile and race, New York

mode-adjusted accessibility to jobs varies notably among the three racial groups in the New York region (figure 6.5). Asians as a group experience higher accessibility than the other two groups. Blacks and whites are similar through most of the distribution, except for a small segment where blacks are advantaged in the medium-to-high range of accessibility.

The case of Los Angeles offers a contrast to that of New York in the comparison of blacks and whites (figure 6.6). In Los Angeles, blacks tend to experience higher accessibility than whites through most of the distribution, a pattern that is substantially different from New York's. The difference may be due in part to a higher degree of centrality among blacks in Los Angeles as compared to New York, which may reflect differences in constrained residential location choices or other historic settlement patterns. The difference in car-ownership rates between blacks and whites also probably plays an important role in explaining the variation found in the figures for New York and Los Angeles. For example, a substantially larger share of African Americans in Los Angeles

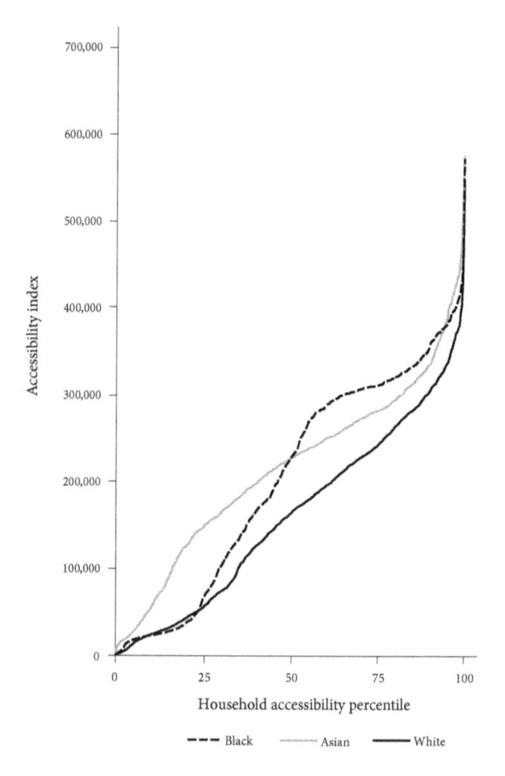

FIGURE 6.6. Mode-adjusted accessibility to work by household accessibility percentile and race, Los Angeles

live in households with vehicles available (figure 6.7). Furthermore, the gap in vehicle availability between blacks and whites is considerably smaller in Los Angeles than in New York. The advantage that black Los Angelinos experience is likely driven in large part by high black vehicle-availability rates, especially if a large share of blacks in the Los Angeles region are living in central locations (figure 6.6).

The accessibility advantage that blacks as a group experience in Los Angeles and New York may run counter to the expectations of some. Indeed, the conventional understanding in the social science literature is that racial minorities are disadvantaged in getting to opportunities in the United States because a growing share of metropolitan destinations are located in distant suburbs, while minorities live near the center.[52] Yet, from an accessibility perspective, central locations position their residents nearer to a wider range of opportunities than anywhere else in a region.[53]

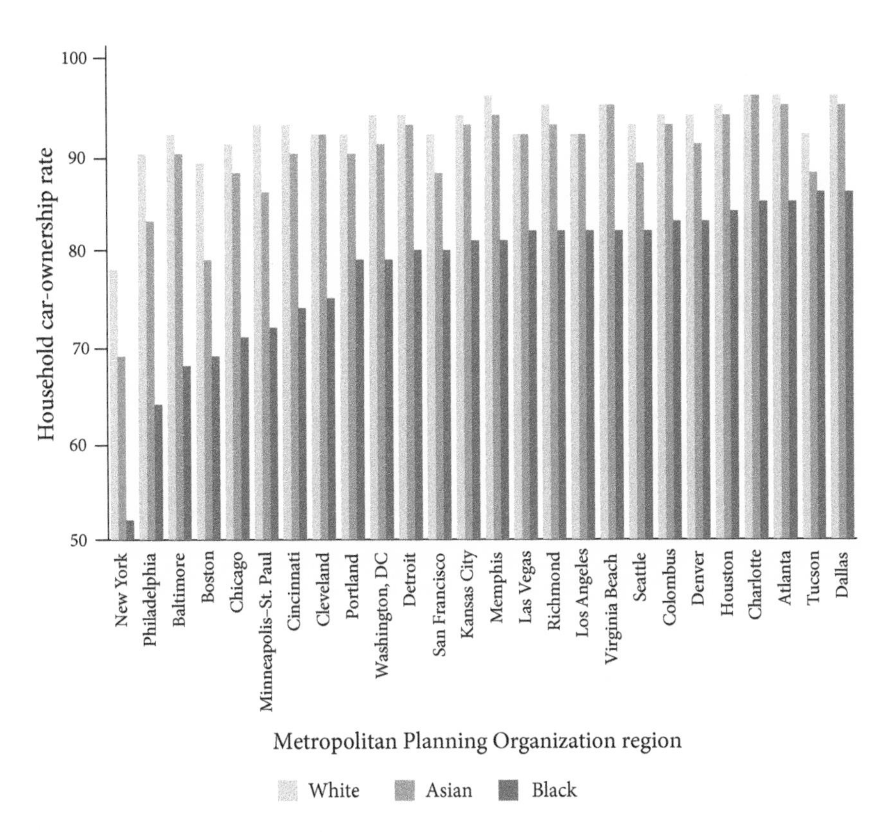

FIGURE 6.7. Intermetropolitan comparison of vehicle-availability rates, by race (sorted in increasing order of African American car ownership)

In this case, the equity dimension of accessibility may be more one of meaning than magnitude, because the accessibility that blacks experience varies for different reasons than for whites. The majority of variation in white accessibility is likely a product of freely chosen residential location, while the majority of the variation in black accessibility may be a product of car availability. Monitoring changes in such variables is important from a public-policy perspective. Whites tend to live disproportionately in low-density communities, and they may be paying a price in the form of low accessibility for doing so, consistent with land-market theories that suggest that some households willingly trade off higher transportation costs in exchange for larger homes and lots at distant locations.[54] Seen in this way, differences in the matter of choice between blacks and whites are important to recognize in evaluating accessibility outcomes.

The finding that blacks are advantaged in accessibility terms in Los Angeles and New York does not extend to all U.S. metropolitan areas. Several metropolitan regions reveal patterns of distinct accessibility disadvantage for blacks. For example, African Americans in Cleveland, Boston, and Philadelphia are severely disadvantaged throughout the entire accessibility distribution (charts not shown for brevity). All three of these regions are places with relatively low vehicle-availability rates among blacks (as shown in figure 6.7), yet other regions with comparably low vehicle-availability rates among blacks, such as Chicago and Minneapolis–St. Paul, do not reveal a similar accessibility disadvantage. These results underscore the importance of analyzing differences among social groups on a region-specific basis by considering the effect of mode simultaneously with the effect of residential location.

Social Equity and Urban-Form Dimensions of Regional Accessibility

Accessibility is inherently linked to places and the people who live in them, and accessibility analysis allows planners to identify which people are disadvantaged in their ability to reach destinations. Where mobility-based analysis leaves a great deal of uncertainty about whether a metric actually indicates disadvantage or not, the meaning of an accessibility metric is consistent and clear because an improvement in the ability to reach destinations is the ultimate goal of the transportation system.

The accessibility shift alone is not enough to advance social equity, but accessibility metrics are essential for informing sound policy choices. The analyses presented in this chapter acknowledge that multiple goals are in competition in decisions about transportation policy, but they show how accessibility metrics, when linked to the people experiencing that accessibility, offer important insights into questions of social equity in transportation. In the case of metropolitan regions of the United States, no other determinant is as important as the availability of a car in achieving high accessibility. Lacking a car represents a severe disadvantage in accessibility nearly everywhere, and elevating social equity in transportation will require tools, such as those presented here, to direct resources to transit-dependent people.

Aside from the central importance of mode, accessibility-based analysis underscores the importance of residential location in determining people's ability to access opportunities. Given a particular distribution of auto and transit accessibility, the location of carless people still shapes the accessibility that people experience in a region. The match between transit-dependent households and transit-rich zones varies substantially between regions. Improving that match opens up possibilities for advancing social equity with tools traditionally outside of the transportation realm, such as land-use regulatory reform and housing policies that make it easier for people to choose to live where transit is plentiful.

7

NONWORK ACCESSIBILITY

Notwithstanding the attention that the commute receives in transportation-planning practice, the work trip accounts for only a minority of travel; trips to or from work count for only 16 percent of daily person trips and 19 percent of distance traveled in the United States.[1] For this reason, job accessibility in isolation would only partially capture people's ability to reach their destinations and thus be an incomplete accessibility assessment. In this chapter, we develop a comprehensive index of nonwork accessibility, apply it to large metropolitan regions in the United States, and compare the distribution of nonwork accessibility to work accessibility.

Nonwork differs from work accessibility in three important ways. First, people's propensity to travel long distances is significantly greater for work trips than for nonwork trips. At 12.2 miles, the average work trip by car in the United States is nearly twice the distance of the average shopping, family, or personal trip.[2] The relative shortness of nonwork trips may result from the closeness of available alternative destinations (which renders longer-distance travel unnecessary), a resistance to traveling longer distances for nonwork purposes, or some combination of the two. In any case, metrics of nonwork accessibility need to take into account people's greater resistance to long-distance travel to nonwork destinations than to workplaces.

Second, there are frequently good substitutes for nonwork destinations. One grocery store or restaurant may be readily patronized instead of another. Some nonwork destinations are not substitutable—there's no good alternative to "Mom's house" and no good short-term alternative to "my school"—but for much nonwork travel, reasonably substitutable destinations are the norm. Indeed, for many nonwork destinations, such as a trip to a park, the trip is only worth making if it is convenient to do so. By contrast, for most workers, "my workplace" has a unique and fixed location as long as they remain in their jobs; even those who work at multiple sites rarely have complete discretion as to where they work on a given day. To be sure, work locations other than one's own remain highly relevant to accessibility. Because U.S. workers change jobs more frequently than they change addresses—the median U.S. resident changes jobs every 4.1 years and changes residences every 6.2 years[3]—locating one's residence with accessibility to significant job concentrations can be an important adaptability strategy. But overall, the substitutability of work locations pales in comparison to people's short-term ability to choose among nonwork locations.

Third, nonwork destinations are far more varied and less standardized than work destinations. While not all jobs are created equal—and not all jobs are relevant to any given individual—job locations may be combined with each other, such that the statement "zone X contains 10,000 jobs" is meaningful. By contrast, a nonwork destination may be a park, a house of worship, the office of a notary public, a supermarket, a regional shopping mall, or the house of a friend. The differing meaning of each of these destinations and their differing capacity to attract trips impede quantification of the concept of nonwork accessibility. Unlike jobs, which can be readily counted, nonwork destinations cannot be meaningfully combined with one another without additional sorting, weighting, or other processing to render them meaningful along a single scale. One consequence of this incommensurability is that nonwork accessibility is frequently defined sectorally as opposed to comprehensively, as when analysts investigate access to food stores or health-care providers.[4]

The Importance of Nonwork Accessibility

Despite the large proportion of nonwork travel in total trip making, work-related accessibility continues to hold an outsized importance to outcomes of interest to transportation and land-use planners, notably residential-location choice, and significant scientific and policy debates surround the relative importance of work and nonwork accessibility. The prevalence of nonwork travel has led some observers, such as Peter Gordon, Ajay Kumar, and Harry Richardson and Kang-Rae Ma and David Banister to argue that the work trip has been overemphasized in transportation and land-use policy and research:

> Accessibility to the workplace is the primary determinant of residential land rents and location patterns in the standard monocentric model

> of urban land use that remains, much to our regret, the centerpiece of urban economic theory. Both the models constructed by transportation analysts and the investment projects and policies implemented by transportation planners are almost always based on the assumption that the work trip is the only trip that counts.[5]

> The recent growth of trips for other purposes may also result in longer commuting because many households may consider trade-offs among a number of possible activities (e.g. families with children may prioritize access to good schools). This means that a policy of jobs-housing balance could cause greater daily travels if the spatial distribution of non-work trips is not jointly considered. Therefore, minimizing the commute should not be a policy target on its own, particularly when non-work trips are increasingly important in the policies aimed at enhancing local and regional accessibilities.[6]

Whether one's concern is improving households' accessibility, reducing vehicle-kilometers traveled, or coping with roadway congestion, the sheer volume of nonwork travel merits policy attention. Yet studies of the impact of work and nonwork accessibility on household residential location—a central outcome of policy concern due to its effects on all the outcomes we have mentioned—suggest that nonwork accessibility shapes people's locational choices only weakly, while workplace access is one of the most influential variables in multivariate models of residential location choice.[7] For example, access to retail, food, and other services was found to have some influence on households' residential choice, but much less than job-accessibility variables.[8] Similarly, a study of relocating residents in the Puget Sound, Washington area, found that workplace accessibility had a significant influence, but the retail-accessibility variable tested had none.[9] Nonwork accessibility may affect residents' everyday lives in important ways, but it does not seem to have a great influence on most people's residential-location choice.

Development of a Nonwork Accessibility Index

The absolute importance of workplace accessibility to residential location choice may not be surprising, but its strong influence relative to the larger share of travel represented by nonwork destinations remains unexplained. In order to explore this issue, a comprehensive metric of nonwork accessibility is required, in contrast to most indicators of nonwork accessibility, which have been specific to sectors such as retail or services.[10] One approach to constructing a general index used people's duration of nonwork activities as an indicator of the attractiveness of nonwork destinations; the more time a person spent at a destination, the more valuable it was considered to be.[11] Since this metric depends on trips actually taken to the specific destinations in question, it is incompatible with the notion of accessibility as potential for interaction. Another approach defined six types of destinations (cultural services, educational services, health services and facilities, sports and recreational facilities, bank branches, and other types of services and facilities) and used cluster and factor analyses to combine these into an accessibility metric.[12] In an accessibility study at the intercity scale, the gross domestic product of each urban agglomeration was used as the indicator of destination attractiveness.[13] Another study implemented Torsten Hägerstrand's space-time accessibility concept using locations of actual activity participation from a household travel survey.[14] But comprehensive, as opposed to sector-specific, assessments of household nonwork accessibility from the perspective of opportunity for interaction are rare.

A general metric of the attractive capacity of nonwork destinations should have the following attributes:

a. It should include the full range of nonwork destinations, rather than limiting itself to a single sector.
b. It should weight nonwork destinations by their relative attractiveness.
c. It should be consistently implementable with available data for multiple metropolitan areas.

Because of the broad range of nonwork destination types, their definition is somewhat more intricate than that of their workplace counterparts. The metric developed here identifies nonwork destinations from a comprehensive database of work destinations. The guiding concept is that job locations can provide an indicator for the location of nonwork destinations, since many nonwork destinations such as shopping or service locations, restaurants, and schools are also the sites of jobs. But whereas every job is relevant to employment accessibility by definition, only some jobs are indicators of nonwork destinations. For example, while both a job in a steel foundry and a job in a supermarket are relevant to the measurement of employment accessibility, only the latter is included as a nonwork destination. Conversely, the implicit assumption is that every significant nonwork destination can be located through the presence of some jobs (or residences) on site. This assumption does not hold universally; for example, most jobs associated with city parks, which are proper nonwork destinations, would be reported with locations at the municipal headquarters, rather than in the park itself.

The attractiveness of a nonwork destination type is gauged here through its nationwide capacity to attract trips. This attractive capacity is based on data from the U.S. National Household Travel Survey (NHTS), which categorizes nonwork travel into thirty-five purposes including shopping, errands, and purchasing goods; meals and other social events; visiting friends and relatives; exercise or sports; and purchasing services (appendix A, table 7a).[15] The distribution of nonwork trips in a given metropolitan area is assumed here to match the U.S. distribution of trips in the NHTS. This assumption is a limitation of the method, but the alternative of relying on household travel surveys from multiple metropolitan planning organizations (MPOs) would impede the consistent comparison of nonwork accessibility between metropolitan areas.

If trip-attraction capacity is the metric of the relative attractiveness of a nonwork destination, then shares of nonwork travel should guide the development of a weighting scheme within the index. For example, if retail accounts for 29 percent of nonwork trips, it should account for 29 percent of the nonwork accessibility index. This method required a correspondence between NHTS destinations and North American Industry Classification System (NAICS) codes, a system used to characterize jobs by industrial sector in the United States. Developing this correspondence required certain modifications. For example, some classifications in NHTS such as "transport someone," "pick up someone," or "drop someone off" are reasonably clear as trip purposes but cannot readily be linked to specific destinations and hence were dropped from the analysis. Other trip-purpose classifications, such as "buy goods: groceries/clothing/hardware store," "medical/dental services," or "get/eat meal" were easily assigned relevant NAICS codes. Some trip purposes (e.g., "go to religious activity" and "attend funeral/wedding") were merged because while the trip purposes were distinct, there was significant overlap in the destinations. Finally, one trip purpose, "visit friends/relatives," had no potential indicator among the employment-based NAICS codes; population, rather than jobs, was used as the indicator for this trip purpose.

In some cases, the analysis excluded business types because they were not relevant to accessibility from one's home zone. For example, hotels and other tourist destinations were excluded since they are usually most relevant when people are not making trips from home. Car dealerships, which account for a significant number of jobs, were excluded since the low frequency with which nonemployees visit these establishments renders them only marginally relevant to home-based accessibility. By contrast, hospitals were included despite their low frequency of visits since these were assumed to have a significant value independent of actual use: their presence provides an assurance that augments one's accessibility even when they are not needed. Dropping and merging categories resulted in a set of twenty-four nonwork-trip categories that could be linked to relevant job or population types (appendix A, table 7a).

With a set of nonwork destination types defined, and the number of jobs or population associated with those destinations identified, the potential attractiveness of a job or person in each category was assessed as the number of nonwork trips that the average job or person from these categories attracted per day. This number was approximated in relation to the number of work trips based on the assumption that each job attracted one work trip per day; this assumption, together with the number of jobs in the region and the share of work trips throughout the United States, formed the basis for estimating the total number of regional nonwork trips.

Finally, nonwork accessibility indicators were calculated following the standard gravity model formulation, with the impedance coefficient adjusted for nonwork travel according to the method described in Levine et al.[16]

The nonwork attractiveness index is calculated in four steps:

1. Estimate total nonwork trips for a region. Using the total number of jobs in a metropolitan region and the ratio of nonwork to work travel, estimate nonwork trips as follows:

$$N_m = J_m \left(\frac{0.823}{0.177} \right) \tag{7.1}$$

   where N_m is the estimated number of nonwork trips for metropolitan region m, J_m is the number of jobs in metropolitan region m, and the fraction in the right-hand term is the ratio of the national share of nonwork to work trips.

2. Distribute the total nonwork trips across the categories of trip purposes. Assuming that the distribution of trip purposes is the same in a metropolitan region as it is nationwide, express this as:

$$N_r = N_m (E_r) \tag{7.2}$$

   where N_r is the number of nonwork trips in a metropolitan region for trip purpose r, and E_r is the share of national nonwork trips for purpose r.

3. Express the nonwork trips by purpose on a per-job basis. The relevant jobs are those in the subset of all jobs that correspond to specific nonwork destination types.

$$\frac{N_r}{J_r} \tag{7.3}$$

J_r is the number of jobs in the metropolitan region associated with trip purpose r. For the trip purpose "visiting friends and relatives," the attractor is not jobs but population.

4. Calculate the nonwork attractiveness of each zone.

$$W_i = \sum_{i=1}^{k}(N_r \div J_r)J_{ir} \tag{7.4}$$

where W_i is the nonwork attractiveness index at zone i, and J_{ir} is the number of jobs in zone i associated with trip purpose r.

This zone-based nonwork attractiveness index—which is generalized over the course of the day, rather than time-of-day specific—then becomes the measure of destination attractiveness used as an input to calculate nonwork accessibility metrics, whether by car, public transport, walking, or cycling. The metric captures an overall average attractiveness level throughout the day. The following sections demonstrate its use in the case of auto accessibility across thirty-eight of the largest metropolitan regions in the United States.

Nonwork Auto Accessibility in Selected U.S. Metropolitan Areas

The index enables a comparison of nonwork accessibility across metropolitan regions based on a wide range of nonwork destinations. Like the metrics of workplace accessibility, these are associated with metropolitan area size, with the more populous metropolitan regions offering greater nonwork accessibility (figure 7.1). Within very large metropolitan areas, Los Angeles's nonwork accessibility exceeds that of New York's up to a point just beyond the median resident (in accessibility terms). That is, for residents at the low end of the accessibility spectrum (generally those living in the farther-out suburbs), Los Angeles residents enjoy greater nonwork accessibility than their New York counterparts. By contrast, at the higher (more central) end of the accessibility spectrum, New York offers greater nonwork accessibility. The New York/Los Angeles gap becomes very large above the 75th percentile, presumably because New York has much larger concentrations of destinations in central locations. Chicago's nonwork accessibility, by contrast, is considerably lower than either New York's or Los Angeles's, in large part because Chicago has relatively slow traffic (as analyzed from the peak-period travel-time matrices supplied by the respective metropolitan planning organizations). Figure 7.1 demonstrates similar comparisons for large, medium, and small metropolitan regions.

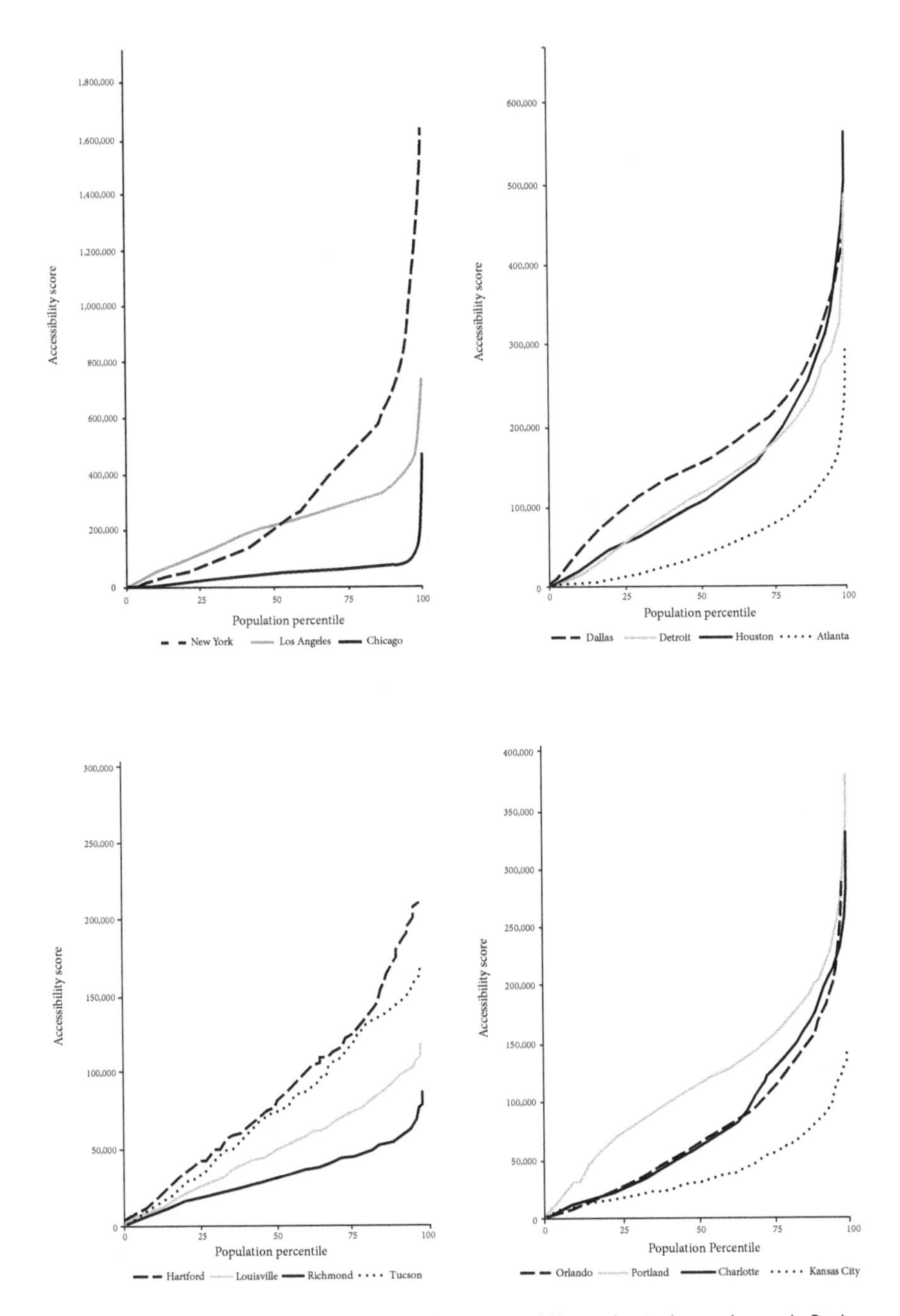

FIGURE 7.1. Nonwork accessibility by automobile, selected very large (>9m), large (5–9m), medium (1.5–5m), and small (1–1.5m) metropolitan planning organization areas (clockwise from upper left)

Nonwork Accessibility's Weak Influence on Residential Location

A comprehensive metric of nonwork accessibility by car can help investigate questions that sectoral models of nonwork accessibility (such as accessibility to retail or health care) are unable to answer. For example, with over 80 percent of travel being to destinations other than work, households might be expected to make their decisions about where to locate largely on the basis of nonwork, rather than work, accessibility. One possible explanation for the relatively weak influence of nonwork accessibility on residential location choice is that nonwork accessibility is more broadly distributed than workplace accessibility. If this were the case, then perhaps households would not need to consider nonwork accessibility much when deciding where to locate; virtually no matter where they live, they would find desirable nonwork destinations within reach. This section uses the comprehensive nonwork accessibility index to test this candidate explanation for the relatively weak influence of nonwork accessibility on where people choose to live.

An investigation into how broadly or narrowly nonwork accessibility is distributed needs to start with an investigation of the distribution of nonwork destinations. The nonwork attractiveness index (W_i), an indicator of those destinations, was nearly 11 percent more dispersed than work destinations. In this, the dispersion of nonwork-destination attractiveness was closer to that of the population's spread-out residential locations than of concentrated metropolitan work destinations, as measured by the standard deviational ellipse.[17] With a greater dispersion of destinations, nonwork accessibility could potentially be more evenly spread than its workplace counterpart.

But while accessibility assessment starts with measures of destination attractiveness (W_i), it also incorporates network travel time and people's willingness to travel for work or nonwork purposes. The evenness of accessibility's distribution across a territory can be measured with the Gini Index, as demonstrated in chapter 6. Recall that a Gini Index value of zero indicates a perfectly even distribution (every zone experiences the same level of accessibility), and a value of one indicates maximum inequality (one zone experiences all the accessibility that a region offers, and all others have none). A larger Gini Index thus indicates that the distribution of nonwork accessibility is more unequal, with accessibility concentrated in a smaller share of the regional territory.

Surprisingly, for most regions, nonwork accessibility tends to be less evenly spread than its workplace counterpart; the median Gini Index across the regions is 0.31 for work and 0.36 for nonwork accessibility (figure 7.2). Since nonwork location attractiveness is more broadly spread than workplace locations, the explanation for this difference lies principally in people's significantly lower willingness to travel long distances to nonwork destinations. The willingness of most people to travel greater distances to work than to nonwork destinations has the effect of distributing workplace accessibility relatively evenly despite the fact that workplaces are more concentrated than residences or nonwork locations. By contrast, the small search radii for nonwork destinations tend to concentrate nonwork accessibility into a more limited set of locations. When people expect grocery stores, restaurants, or houses of worship to be close by, their absence from the neighborhood can create a significant local deficit in nonwork accessibility.

For this reason, the relatively low influence of nonwork accessibility over people's residential-location choices is apparently not explained by a broader distribution of nonwork accessibility across metropolitan regions. The explanation would need to be found elsewhere—for example, in the greater flexibility in choosing where and when to travel in meeting nonwork needs, or in the inability of some households to opt into high-nonwork-accessibility locations because of discrimination, lack of affordability, or other barriers. Land-use regulation that constrains housing supplies in areas of high nonwork accessibility could

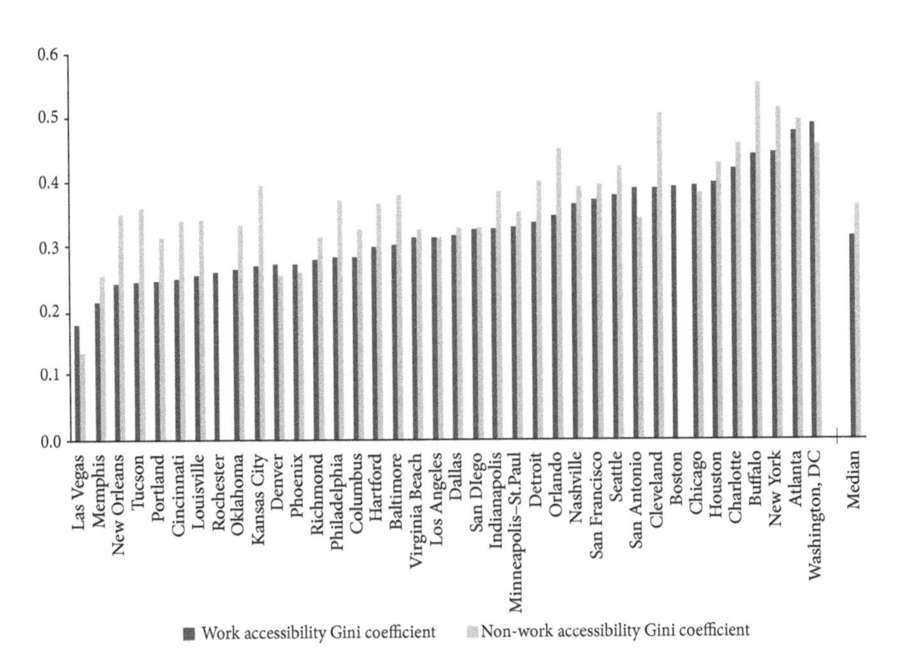

FIGURE 7.2. Work and nonwork auto accessibility Gini Index by metropolitan planning organization area, 2007 (rank order by work accessibility Gini coefficient)

constitute such a barrier, as could a paucity of high-nonwork-accessibility neighborhoods that satisfy potential residents' other desired characteristics, such as safety, school quality, or high-quality municipal services.[18]

In general, larger and more centralized regions tend to have more spatially unequal distributions of nonwork accessibility overall: Washington, DC, New York, Chicago, and Boston are all at the high (spatially unequal) end of the Gini scale. The more even distribution of accessibility occurs in less centralized regions, such as Los Angeles or Phoenix, and smaller regions such as Las Vegas or Tucson. A broad spread of destinations throughout the region tends to reduce intraregional gaps in accessibility by avoiding concentrations that particularly benefit certain zones. Small regions frequently offer all households the ability to reach all other corners of the area in a relatively short time, thus reducing the inequality of accessibility between neighborhoods.

The Continuing Significance of Nonwork Accessibility

Accessibility to destinations other than work has been measured in a number of ways, generally in a sector-by-sector fashion. In this chapter, we describe a method for gauging nonwork accessibility that spans the range of nonwork destinations, as opposed to focusing on a preselected subset. Since the meaning and importance of nonwork destinations vary broadly, the nonwork accessibility index developed here weights nonwork destinations by their average trip-drawing potential. The method complements the more established methods of gauging work accessibility in providing a comprehensive picture of people's ability to reach their destinations and the distribution of this capacity by location and population group.

The limited influence of nonwork accessibility over residential-location choice does not mean that nonwork accessibility is irrelevant to transportation and land-use planning. Quite the opposite: poor access to healthy food degrades nutrition, poor access to schools impedes education, and poor access to medical care hinders the treatment and prevention of disease. All these concerns stem from people's inability to reach places and ought to be central to the professions of transportation and land-use planning.[19] But nonwork accessibility seems to have less capacity than work accessibility to influence residential-location choice, a factor that shapes important outcomes at both the individual and metropolitan scales.

Nevertheless, nonwork accessibility is distributed more unequally than work accessibility in most metropolitan areas; people's desire to have nonwork destinations close at hand means they experience the deficiency of convenient nonwork destinations acutely even though nonwork destinations actually have a wider geographic distribution. The distribution of nonwork accessibility remains an important criterion for transportation and land-use planning success simply because it constitutes the vast majority of trips and distances traveled, and because people's ability to reach meaningful destinations for education, medical care, recreation, and the like is central to their well-being. Comprehensive and broadly applicable metrics, such as the one developed here, can help assess outcomes for both transportation and land-use planning.

Conclusion

ENVISIONING THE ACCESSIBILITY SHIFT

Three logics contend for status as transportation planning's conceptual core: mobility, vehicle-kilometers-traveled (VKT) reduction, and accessibility. The transportation-planning field began in the first half of the twentieth century with a mobility orientation. By the end of the century, many planners and researchers had shifted to VKT reductions as the implicit lodestone of progressive action in transportation and land use, a goal that, by the twenty-first century, made its way into some formal policies—though the mobility paradigm remained dominant overall. We have argued for a logic distinct from both of these: an accessibility shift to align transportation and land-use planning with transportation's core purpose.

Of the three logics, only accessibility is invisible. Immobility is evident to all in the form of traffic jams or slow public transport. High VKT, the target of driving-reduction efforts, is observable from evidence including travel surveys or GPS-based data collection on vehicle movements. By contrast, accessibility is a potential for interaction, present even when that potential is not translated into observable travel behavior. Its very invisibility has impeded people's comprehension and hence their acceptance of accessibility as something real.

In addition to their visibility, mobility and VKT reduction share an important attribute that aligns them with the traditional inclinations of planners and planning institutions. Following prescriptions emanating from welfare economics, a policy-analytic tradition common in the United States and elsewhere generally advises the policy maker to rely on the efficient functioning of the free market.

Governments intervene, within this framework, when inefficiencies arise in the form of market failures, especially negative externalities. By broad consensus, governments exist to prevent or mitigate harm to their citizens; in this way a negative externality, if large enough, constitutes a rationale for governmental intervention.[1] Congestion, the prime target of mobility-based planning, is one such externality. The air toxics, regional pollution, and greenhouse-gas emissions associated with driving—and the prime targets of VKT-reduction efforts—are another. For these reasons, planning for either mobility or VKT reduction enjoys a legitimacy rooted in a broadly shared view of the proper role of government. Limiting action to reduction of harms, such as those associated with congestion or vehicular emissions, can help maintain the appearance of restraint or even neutrality of governmental intervention into the workings of the market.

By contrast, accessibility planning, which is fundamentally benefit seeking rather than harm avoiding, is mismatched with the negative-externality framework. This helps explain why many seek to justify accessibility planning via the instrumental view (see chapter 1): accessibility can be recognized through the negative-externality lens if it proves itself in VKT-reduction (or even congestion-mitigation) terms. Yet this tunnel vision comes at the cost of blindness to accessibility's other benefits.

Notwithstanding the challenges it faces in the form of invisibility, accessibility is the only reliable indicator, among the three contenders, of the benefits offered by transportation. This renders both mobility and VKT reduction inadequate as transportation planning's central logic, an inadequacy that can lead to perverse outcomes. For example, a household experiencing high congestion could be a low- or a high-accessibility household; the latter would be able to interact with many destinations, notwithstanding its low mobility.[2] Fighting congestion for the benefit of such households would result in resources poorly spent, especially considering the potential for the fight itself to degrade accessibility through an induced spread of origins and destinations.

Despite accessibility's inherent value, it can be hard to imagine it playing a major role in guiding planning of the built environment. It may be that planning is relatively insensitive to analytical frameworks and evaluation metrics and proceeds instead according to raw political imperatives. It may be that mobility planning is not really a function of explicit policy choices but the spontaneous and inevitable product of people's immutable interests and desires. Or it may be that planning interventions are not ultimately consequential, and the market wins out in any case.

We have argued against all three of these positions in favor of the idea that how planning defines and measures success is highly consequential. The existing mobility paradigm molds transportation and land-use planning at multiple levels and geographic scales and demonstrably shapes metropolitan development. This power suggests that the accessibility shift similarly holds great potential for altering decisions and ultimately the built environment.[3]

Mobility and VKT after the Accessibility Shift

Though neither mobility nor VKT reduction can capture the benefits of a transportation/land-use system, both are still needed inputs into planning decisions. Mobility analysis is an essential component of, and hence input into, accessibility analysis. Ostensible accessibility analyses that fail to account for mobility devolve into analyses of simple proximity. For this reason, mobility analyses would not be eliminated by the accessibility shift but would be combined with information on land use to produce a meaningful investigation of transportation's benefits.

VKT analyses are similarly needed after the accessibility shift. While accessibility is a metric of transportation's benefits, it does not account for transportation's harms, notably greenhouse-gas emissions and other pollutants. In many cases, accessibility-improving strategies will also reduce environmental externalities, but this relationship is not assured (see chapter 1). A separate accounting for transportation's harms, such as that offered by VKT, is necessary, but a harm-avoidance strategy must not be taken as equivalent to a benefit-seeking strategy.

Accessibility and VKT-reduction goals are, however, interrelated as rationales for planning action. Transportation and land-use planners are keenly interested in land-use approaches to transportation issues, specifically the potential of compact and mixed land uses as a strategy to reduce VKT. This strategy faces a distinct limitation in the form of the slow pace of change of urban form. Elements of the built environment are long-lived, and the change in any given year will be small as a proportion of the established stock. Given these challenges, the aspiration for measurable regional travel-behavior shifts often seems unreachable in the short term. Moreover, marginal change in the form of a walkable neighborhood in an auto-oriented region will likely support some car-use reduction, but since much travel behavior is a function of the broader metropolitan environment rather than of an immediate neighborhood, this effect would be limited.

Large-scale change in development patterns may promise more thoroughgoing reductions in people's use of cars, but promoting current policy on the basis of potential long-run gains will always face an uphill political battle. Given this challenge, accessibility can provide the immediate, short-term payoff needed to demonstrate the benefits of many types of compact development. A new neighborhood with easy access to work and nonwork destinations offers its residents notable accessibility benefits, even if those benefits do not materialize as significant VKT reductions in the short term. Demonstrable short-term accessibility gains can obviate the need for immediate proof of benefit in VKT-reduction terms, allowing the travel-behavior benefits of compact development to emerge over the longer term.

The implications of the accessibility shift for planning practice may be farther reaching still. Planners understandably seek to demonstrate that public-transport deployment and land-use approaches of compact, walkable, transit-friendly urbanism lead to desirable environmental outcomes by reducing VKT. Yet gains in air quality since the 1970s have largely stemmed from technological advances such as the catalytic converter, rather than from planners' efforts at moderating VKT.[4] Continued technological improvement in the automobile may undermine the ability of transportation and land-use planners to rely on atmospheric environmental concerns as a core basis for their actions. For example, a future of sustainably powered electric vehicles, whether conventional or self-driving, individually owned or shared, would greatly reduce the usefulness of VKT as a metric of transportation's harms. Other environmental harms related to the auto would still be relevant, to be sure. The automobile would still impose environmental harms in the form of non-point-source water pollution because of the impervious surfaces needed to support an auto-intensive transportation and land-use system. Low-density urban forms associated with the automobile would still lead to habitat depletion, and transportation-system noise would still affect neighborhoods near major roads. But the elimination of atmosphere-borne harms in the form of air toxics, regional air pollution, or greenhouse gases, an unmistakable boon to global sustainability, would weaken any transportation-planning rationale grounded in VKT reduction. The claim that land-use patterns are an important environmental-policy instrument for their capacity to reduce VKT may well be undermined by future technological improvements.

By contrast, the accessibility paradigm in transportation and land-use planning will remain a robust basis for action, as sound after the welcome technological change as before. A capacity to interact with destinations will still be the service that people seek from their transportation systems, and that capacity will still be delivered through some combination of mobility, proximity, and connectivity. Land-use patterns, which determine the degree of origin-destination proximity, will continue to be relevant to the delivery of accessibility.[5] Planning will continue to shape transportation systems, through investment, and land-use systems, through regulation. Then, as now, accessibility can serve as the logical foundation of the transportation/land-use planning enterprise.

Accessibility as Transportation/ Land-Use Integration

Yet a disjuncture between the planning of transportation and the planning of land use impedes the pursuit of accessibility. Across multiple contexts worldwide, the planning field aspires to integrate the two realms, which are commonly planned for in relative isolation from each other. In most countries, the bodies that plan for highways and public transport are distinct in their institutional setting, professional logic, political imperatives, and geographic scale from those that plan for the physical development of cities and regions. Disconnected planning frequently produces unsatisfactory results. Disjuncture between the two realms abounds, for example, where land-use authorities permit development in zones of poor transportation accessibility with no plans for connecting these areas with the existing transportation system, or where transportation planning expands roadway access to areas ostensibly slated for preservation. But like many watchwords enjoying near-universal acceptance, "transportation/land-use integration" can mean many things.

One dimension of the potential integration is institutional.[6] According to this thinking, if transportation and land-use planners interacted more, or more meaningfully, or with better sharing of information and assumptions, they would plan at cross purposes less frequently. The two systems are so entirely interrelated, in this account, that the separation of realms is inherently damaging. The barrier to integration of the two realms is often seen in terms of geographic scale; whereas in many countries the municipality is the primary locus of land-use planning, transportation planning occurs at the larger level of the metropolitan region, province, district, or state. Aligning the geographic scale of the two endeavors is key to rationalizing their respective planning processes. Even better, presumably, would be if transportation and land-use were planned by the same institutions. Although this version of the integration of transportation and land-use planning has much to offer, a purely institutional definition of transportation/land-use integration without guiding principles leaves unanswered the question: To what ends is such integration to be directed?

One common substantive definition of transportation/land-use integration is represented by requirements that land development not be allowed to occur unless sufficient transportation capacity, often interpreted as roadway capacity, is provided. Since these requirements are oriented toward guaranteeing sufficient infrastructure to maintain automotive level of service, they amount to a mobility-based transportation and land-use integration. Thus, while these requirements surely represent a substantive integration of transportation and land-use planning, they can impede compact development, and as a consequence, metropolitan accessibility (see chapters 3 and 4). These outcomes would seem antithetical to the goals of many proponents of transportation and land-use integration.

The accessibility shift provides a sounder answer to the question "Integration for what purpose?" By incorporating attributes of both the transportation and land-use systems, the accessibility concept inherently integrates the two realms. Advocates for transportation/land-use integration tend to support land development in areas of high transportation accessibility, transportation investments that serve already built-up areas, and improved spatial match between jobs and housing. By shifting the definition of transportation success, the accessibility shift can ease the way toward these policy reforms. Unlike the institutionally defined version, it specifies the ends to which the endeavor is to be directed, and unlike the version that insists on sufficient roadway capacity prior to land development, it serves the core purpose of transportation.

Mobility Planning Ideology

Envisioning the accessibility shift depends in part on revealing the current default: mobility-based transportation and land-use planning. Where accessibility can be invisible because of its intangibility, its mobility counterpart can be obscured by its ubiquity. Mobility thinking and formal mobility analyses pervade transportation and land-use planning from the local level, as when land-use decisions are guided by their anticipated impact on car traffic, to the regional scale, as when metropolitan areas compare themselves with their peers based on congestion rankings.[7] Mobility-driven decisions are evident wherever forecasts of roadway congestion are used to guide investment in transportation infrastructure. The methods of mobility appear in the manuals of professional organizations such as the *Highway Capacity Manual*, the "Green Book" of the American Association of State Highway and Transportation Officials, and the *Traffic Engineering Handbook* of the Institute of Transportation Engineers.[8]

Mobility planning is even enshrined in law. In the United States, it has been reinforced at the national level at least since the Federal Highway Act of 1963, which mandated a "continuing, comprehensive, and cooperative" transportation-planning process as a condition of federal funding. The mandate, while not specific about methods, served to reinforce a mobility-oriented transportation-planning system as the foundation of regional transportation planning, since travel-demand modeling was the primary transportation-planning tool available at the time. More recently, guidelines based on Moving Ahead for Progress in the 21st Century Act (MAP-21), a federal transportation law in the United States, mandated that state departments of transportation and metropolitan planning organizations monitor traffic congestion, excessive delay, and travel-time reliability as three different dimensions of mobility measurement. At the local level, municipal ordinances in cities nationwide mandate land-use planning aimed at maintaining highway level of service (text box C.1), often known as "concurrency."

Mobility-based planning exists not only in the application of formal analytical models and metrics, but also in the assumptions of participants in the transportation and land-use planning process, from engaged citizens to planners, engineers, and decision makers. Even in the absence of quantitative forecasts of congestion, popular worry about traffic can doom land-development projects; public discourse frequently conveys an unspoken presumption that roadway congestion is the principal lens through which to scrutinize a proposed project's desirability. This is expressed in the near-universal complaint—which is sometimes a stalking horse for class or race concerns—that new development will bring more traffic to a neighborhood.[9]

Mobility planning's entrenchment in the transportation and land-use realms can be daunting to the would-be accessibility reformer. But embedded in its deep institutionalization lies a potential key to systemic change. Transportation and land-use planners, seeking to ground their policies and investments in quantitative analysis of transportation and land-use conditions, know well how to incorporate metrics and the product of models into their decision making.[10] The accessibility shift, while a potentially radical conceptual change, is methodologically no more than grafting a different set of metrics and methods onto existing professional branches. The data needed are largely those in current use, and in any case, standard mobility analyses remain as needed inputs into accessibility-based decision making.

Notwithstanding its ubiquity, mobility-based planning is anything but a neutral default approach, and it is not merely the technical aggregation of preferences of individual consumers or voters. It is the inscription of a particular ideology into the built environment. The statement of the Institute of Transportation Engineers that a transportation-impact analysis "should not include political views or statements, nor should it take an advocacy position" is a logical impossibility, since the analysis itself is grounded in a normative commitment to mobility (nearly always interpreted as vehicular mobility) as the proper goal of transportation and land-use planning.[11] The accessibility shift does not escape the normative nature of mobility planning; no notion of the good in guiding plans and policies for the built environment could.[12] Instead, it aligns planning ideology and practice with what the planning field already accepts about the purpose of transportation.

Text box C.1: Examples of laws and ordinances mandating mobility-based planning

City of Snoqualmie, Washington: Development shall be prohibited, and all applications therefore shall be denied, if the development would cause the level of service on a transportation facility to decline below adopted levels of service standards, unless the transportation improvements or strategies to accommodate the new development are made concurrent with the development, subject to the provisions of subsection D of this section.[13]

Wake County, North Carolina: The traffic impact analysis must demonstrate that the proposed development would not cause build year peak hour levels of service on any arterial or collector road or intersection within the analysis area to fall below Level of Service (LOS) "D" (as defined by the latest edition of the highway capacity manual) or, where the existing level of service is already LOS "E", that the proposed development would not cause the LOS to fall to the next lower letter grade. If the road segment or intersection is already LOS "F", the traffic impact analysis must demonstrate that the proposed development, with any proposed improvements, would not cause buildout year peak hour operation to degrade more than 5% of the total delay on any intersection approach.[14]

Cities of Pocatello and Chubbuck and Bannock County (Idaho): Impact Studies evaluate the potential impact of a specific development on the existing system. Acceptable level of service standards for each intersection evaluated shall be in accordance with [the following principles]:

- When the LOS without development is LOS A, B, and C, the minimum acceptable projected LOS shall be LOS C for all movement within a specific intersection.
- When the LOS without development is LOS D, E, or F the minimum acceptable projected LOS shall be equal to the LOS without development.[15]

The Multiple Faces of Mobility Planning

Mobility-based planning shapes public decision making both formally and informally and operates on both transportation and land use. In many cases, mobility planning influences or even determines built-environment outcomes; in others, planning strategies geared to the pedestrian, cyclist, or transit user are adopted after overcoming the resistance of operating against the mobility grain. And mobility planning does not operate in isolation but rather interacts with broader mobility-based thinking and local politics to produce planning decisions.

This section illustrates typical avenues of influence of mobility-based planning through three case studies of transportation and land-use planning, located in Albuquerque; Camden County, Missouri; and San Francisco. In combination, they exemplify the influence of mobility on both transportation and land-use planning, the role of formal processes and informal norms, and the obstacles mobility planning can place in the path of any transportation or land-use innovation that deviates significantly from accepted practice.

Formal Mobility Modeling

Around 2000, Paseo del Norte, a major east-west route in Albuquerque, was serving as an important connection between the rapidly growing residential development west of the Rio Grande and employment and activity along Interstate 25.[16] Mitigating traffic congestion along Paseo del Norte at the I-25 interchange became a priority in the region, and the resulting project represents a textbook example of the application of mobility methods to transportation planning. Paseo del Norte was particularly important because it contained one of seven bridges that crossed the Rio Grande to connect the west side of Albuquerque with Sandoval County. The City of Albuquerque, the New Mexico Department of Transportation (NMDOT), the Mid-Region Council of Governments (MRCOG), and the Federal Highway Administration (FHWA) considered a $350 million plan, ultimately settling on a scaled-down $93 million plan to reconstruct the interchange.

Several analyses conducted at various stages of planning illustrate the prevalence of mobility-based metrics in justifying a project. Before securing funding, the city applied for a Transportation Investment Generating Economic Recovery (TIGER) grant to fund the project. MRCOG conducted an Economic Impact Report, published in November 2011, as part of the cost-benefit analysis required for these applications. The analysis modeled average speed, average trip time, vehicle miles traveled, vehicle hours of travel, and vehicle hours of delay for the build and no-build alternatives to evaluate 2015, 2025, and 2035 outcomes.

In describing the purpose of the project, the required report referenced a Congestion Management Process conducted by MRCOG, which summarized statistics on the thirty most congested corridors in Albuquerque based on volume-to-capacity ratio, speed differentials, and crash rates.[17] The Congestion Management Process ranked Paseo Del Norte as the thirteenth overall most congested corridor, specifically noting the high congestion during rush hour, during which the corridor carried sixty to eighty thousand vehicles per day. The analyses used Level of Service (LOS scores A through F) to compare traffic volumes at peak hour in 2035 with the no-build and the preferred projects.[18] The report demonstrated that the existing facility was not adequate for the projected travel demand. The project was added to the Transportation Improvement Program (TIP), and the proposed project was deemed to have a "smaller environmental footprint" than its alternatives.[19]

As with any formal planning analysis, the political assessment of needs was less formal than the quantitative modeling, beginning with a view developed between the various players as far back as 2001 that capacity needed to be expanded to accommodate current and future vehicular flow. By 2011 the project had been languishing due to funding difficulties, and according to one former politician, a politically influential developer who stood to gain from the project was instrumental in its resurrection. The technocratic elements of travel-demand modeling and LOS forecasting were all present but came after political processes determining project need. Yet the political actors were aware of, and surely influenced by, the modes of formal evaluation that would be required if the project was to reach completion.

Accessibility-oriented thinking was absent, save one reference in a 2011 Economic Impact Report arguing that when the network functions more efficiently, "origins and destinations become closer to one another in terms of time and cost."[20] The statement is accurate if one holds locations of land uses constant; by contrast, a full accessibility analysis models land-use response to the major transportation change before any assessment of the time and cost needed to overcome origin-destination separation (see chapter 3). Neither this kind of land-use modeling nor land-use approaches in general played a significant role in this mobility-based evaluation.

Informal Mobility Thinking in Land-Use Planning

Mobility-oriented thinking can shape outcomes in land development as well as transportation even when formal models are not deployed. This was the case in a planning controversy in Camden County, Missouri, a popular summer-vacation destination between St. Louis and Kansas City, where, in 2007, a land developer sought to develop one hundred and eighty condominiums on 4.5 lakefront hectares, which included both low-density residential and general commercial zoning. The project would have required a conditional-use permit because the proposal deviated from the parcels' zoning designation, a permit that was denied first by the Planning and Zoning Commission and later on appeal by the County Board of Adjustments. The developer sued based on the claim that the proposed uses were less intense than the uses for which the land was zoned; the suit lost at both the trial and appellate courts with a finding that the municipality could legitimately rely on the testimony of affected neighbors over that of expert witnesses in denying the conditional-use permit.

Both those who argued for and those who were against the development at the Board of Adjustment's public hearing relied on mobility-based arguments to bolster their positions. The developer's attorney maintained that "there would not be much increased traffic flow in the area based on the nature of condominium units and the fact that they are vacant a great deal of the time."[21] An engineer, a real-estate agent, and an appraiser used similar arguments to support the development. Neighbors opposed to the development expressed concerns including increased traffic. One neighbor asked, "How in the world is a 130 condo unit going to help anything as far as lessening congestion of traffic on the roads?"[22]

In fact, traffic-impact analysis reported that any congestion effects of the proposed development were not particularly severe. Although it would create some traffic impacts, the developer could have alleviated these impacts by redesigning the access points. However, the Planning and Zoning Commission denied the initial conditional-use-permit application, citing traffic concerns supported by mobility-related zoning considerations in three of its nine criteria. After the legal appeals concluded, the entire site was downzoned to low-density residential, presumably to forestall similar proposals and controversies in the future. Ultimately, given the discretionary nature of the conditional-use-permit application—the County Board of Adjustments was not required to grant the requested permit—the denial was a fairly straightforward matter, even though the traffic analysis failed to back up the informal mobility worries of neighbors and other stakeholders. The accessibility point of view, that is, an assessment of whether the condominiums would have increased people's ability to reach valued destinations, was absent from the development-review discourse on both the prodevelopment and antidevelopment sides.

Overcoming the Obstacles of Mobility Planning

The Bicycle Plan of the City of San Francisco, released in 2005, aimed to increase space on city streets for bicycles in order to enhance the safety and amenability of cycling and thereby to increase the number of city residents who would choose cycling. This effort needed to overcome obstacles sourced in mobility-based planning in order to do so. At the time, the California Environmental Quality Act (CEQA) categorized roadway congestion as an adverse environmental impact, which triggered requirements of the project's traffic impacts. Projects such as the Bicycle Plan, which can potentially increase congestion by restricting the space available for vehicles in favor of bicycles, needed to treat any slowing of car traffic as an environmental harm regardless of the positive environmental (and accessibility) impacts of improved bicycle infrastructure. Local residents sued the city in 2006 for neglecting to conduct a full environmental-impact report (EIR) on the 2005 Bicycle Plan, as per CEQA requirements. The court required the City to conduct the EIR and placed an injunction on the plan until it was completed. In 2010, after the city created the 2009 Bicycle Plan Final EIR, which included a full study of the plan's environmental impacts, the court lifted the injunction.

The Bicycle Plan received support from a number of stakeholders, including Mayor Gavin Newsom, who, after the injunction was placed on the plan, said, "Better, safer bicycling is good for San Franciscans and good for our environment. I remain committed to realizing our goal of a citywide bicycle network that unites the current patchwork of bike lanes into a unified, comprehensive system."[23] Despite this support, the potential "environmental" impacts caused by limiting vehicle mobility slowed the enactment of the plan because it triggered the EIR requirement.

The EIR that led to the lifting of the injunction assessed roadway LOS impacts in accordance with CEQA requirements.[24] The LOS evaluations of the overall goals and policies, and minor and long-term improvements, found several potentially significant and unavoidable impacts. For example, the long-term improvements, which included "area-specific changes in circulation patterns" and large development projects, contained the following four potentially significant and unavoidable impacts: "(a) a potential to increase traffic delays in some areas of the City, (b) a potential to cause a significant adverse impact to intersection levels-of-service; (c) a potential to slow transit vehicle movement in some locations; and (d) a potential to eliminate some curb space, currently used for passenger loading/unloading or commercial loading/unloading. All these potential impacts were identified as being potentially significant and unavoidable."[25]

Ultimately, the injunction was lifted, and the San Francisco Bicycle Plan was adopted. But with traffic congestion framed as an environmental harm per se, no provision allowed for its trade-off against potential accessibility benefits. Effectively, promoters of the Bicycle Plan were put in the position of arguing that impacts on car traffic are not overly onerous, or alternatively that options that could mitigate negative LOS impacts were not feasible. A plan designed to encourage an alternative to the car had to explain—and even justify—itself in vehicular-mobility terms.

Accessibility-Based Reform Efforts

Common to the three cases we have mentioned is an immediacy of mobility thinking and metrics to applied planning practice. Choices in either transportation or land use needed to be made, and mobility-based evaluation was one of the first tools that planners deployed to support those decisions. At times, outcomes were shaped by mobility-based thinking, which established the frame of reference of interested parties even before any formal analysis took place.

By contrast, accessibility-based evaluation is frequently removed from day-to-day decision making. Planners and policy makers often view accessibility-based evaluation as merely advisory, or as a visioning exercise at the regional level. These are vital planning functions to be sure, but they have only indirect links to decisions that shape transportation and land-use outcomes. Regional accessibility-driven visions would face obstacles in simply making the leap to on-the-ground implementation, but with mobility-based evaluation firmly ensconced at operative levels, accessibility-related action struggles doubly to find its footing.[26]

Nevertheless, regional planning is starting to incorporate accessibility-based evaluation, an activity formerly dominated by mobility thinking alone, and the process of developing shared visions matters even where implementation is flawed. However, accessibility concepts must clear several hurdles before they can effectively shape regional futures. First, accessibility must be properly conceived as a concept distinct from mobility; frequently, accessibility remains an undefined buzzword, or accessibility and mobility are fused into a single, undifferentiated whole.[27] Second, to be of practical use, accessibility must be operationalized with conceptually sound measures that can influence decision making; often the conceptualization of accessibility invokes a generalized notion of "access" to the transportation system rather than a sharper focus on access to destinations.[28] Third, plans must specify explicit methods for incorporating accessibility into implementation, such as via project selection or development review. So while the trend toward greater integration of accessibility into regional planning efforts is a welcome beginning, much work remains before the accessibility shift is realized. The next section uses instances of accessibility-based planning at the regional level in the United States to illustrate both progress and gaps in accessibility planning; similar themes emerge from international analyses as well.[29]

Accessibility in the Shadows of Mobility

Federal guidance in the United States to address "accessibility and mobility" in the transportation planning process, without accompanying distinctions between the concepts, has led to many planning efforts that either fuse accessibility and mobility or confuse their relationship to each other. The Puget Sound Regional Council's Transportation 2040 Plan is one such example.[30] In some locations within the plan, accessibility is defined as access to destinations, as when accessibility is identified as a "measure of the ability or ease to travel among various origins and destinations."[31] In other cases, accessibility is defined as access to the transportation system itself, as in the following statements: "infrastructure that provides full accessibility to transit facilities, such as sidewalks, can broaden the transit market share for all users of public transportation" and "mixed use developments can build ferry ridership and increase accessibility to passenger ferry services."[32] Conceived of in the latter fashion, accessibility serves mobility (by enabling people to use the transportation system), rather than the reverse.

This framing is consistent with the dominant mobility theme of the plan. The plan's key strategies are (1) improving mobility, (2) protecting and enhancing the environment, and (3) identifying sustainable funding. In a similar manner, the plan articulates its core purpose as follows: "Transportation 2040 was developed to provide accessible, affordable, and convenient mobility to all people in the region. Transportation 2040 is about ensuring that everyone has access to goods, services and jobs."[33] Mobility and accessibility are both recognized, but mobility remains the central organizing concept throughout. Performance-measure selection reinforces the impression of accessibility's subservience in the plan, with mobility metrics prominent and accessibility indicators absent. The outcomes identified for monitoring and evaluating plan implementation include roadway speeds, average per-capita travel time, daily per-capita traffic congestion, and daily person trips per mode. Roadway speeds and per-capita traffic congestion are strictly mobility-oriented performance measures. Average per-capita travel time is loosely related to accessibility but is a measure of observed travel behavior, and not the opportunity to reach destinations per se. For this reason, it fails to capture the value offered by the transportation/land-use system. Standard accessibility metrics such as cumulative opportunity, gravity, or utility-based measures are not included as performance measures within the final planning document. Furthermore, the monitoring and implementation section of the plan calls for the measurement of "accessibility to transportation choices" but not accessibility to destinations. In short, although ideas about accessibility are present within the plan, its overall thrust is about improving mobility, and the role of accessibility is primarily conceived of as access to the various components of the transportation system itself in order to enable such mobility choices.

Accessibility as a Bit Player

In other plans, accessibility is consistently operationalized as access to destinations rather than access to the transportation system itself. Nevertheless, when accessibility plays only a small role among many competing criteria and goals, its effect can be diluted. Unless accessibility constitutes transportation planning's mainstream, it can be lost in a deluge of other planning goals.

The Chicago region embarked on a major regional planning effort in 2008 known as the "GO TO 2040" plan.[34] As part of this regional-planning effort, three major scenarios were developed in consultation with the public and stakeholders. The "reinvest" scenario aimed to target future growth within existing communities and to channel investment into the reconstruction of existing infrastructure. The "preserve" scenario strove to preserve the existing character of built communities and to minimize disruption of existing land-use patterns by avoiding obtrusive transportation projects, focusing instead on small-scale, localized transportation improvements. The "innovate" scenario sought to continue outward metropolitan expansion while also incorporating energy-efficient technologies to mitigate the impacts of this outward growth.

Each of the regional scenarios was fully developed as a future transportation and land-use plan, together with a suite of accompanying implementation policies. These alternative futures were then evaluated against multiple metrics of future regional performance including land consumption, percent of infill development, open-space preservation, water use, land covered by impervious surfaces, energy consumption, congestion, mode share, travel times, accessibility to jobs by auto and transit, future industry mix, and infrastructure costs. This analysis of future scenario performance was presented to stakeholders and the public for further discussion and refinement. The process led to the selection of a preferred regional scenario, which incorporated elements of each of the other scenarios.

The preferred regional scenario and the scenario analysis process were used to identify a series of policy priorities, grouped into four major themes: livable communities, regional mobility, human capital, and efficient governance. Nested within these four major themes were twelve policy priorities:

1. Achieve greater livability through land use and housing.
2. Manage and conserve water and energy resources.

3. Expand and improve parks and open space.
4. Promote sustainable local food.
5. Invest strategically in transportation.
6. Increase commitment to public transit.
7. Create a more efficient freight network.
8. Improve education and workforce development.
9. Support economic innovation.
10. Reform state and local tax policy.
11. Improve access to information.
12. Pursue coordinated investments.

Accessibility played a relatively minor role in the plan-development process, and it was not a major driver in the policy development for the overall GO TO 2040 Plan. Accessibility metrics informed the performance evaluation of alternative scenarios and played a behind-the-scenes role in the development of the preferred scenario, but the accessibility concept was not present within the four major themes or the twelve policy priorities. The Executive Summary explaining these regional policy priorities mentioned the word "accessibility" only once, as follows: "First, the region should provide more parks in developed areas to increase park accessibility and equity."[35] The behind-the-scenes role of accessibility within the planning process offered little opportunity for accessibility-based thinking to shape the implementation of future transportation and land-use decisions. Indeed, the plan explicitly noted that land-use decisions would be left up to local authorities and exhorted them to strive for better transportation/land-use integration without providing a framework for what such better integration would look like.

When accessibility is considered as just one transportation-related criterion out of many, its centrality as the primary measure of benefit provided by the transportation/land-use system is neglected, and attention to its nuts-and-bolts implementation is typically absent. In regional-planning efforts such as this one, accessibility is nominally present, but its potential as a means of guiding the development of the transportation/land-use system remains unrealized.

Accessibility Integrated into Project Selection

When plans identify specific links to decision making, the accessibility shift can influence plan implementation as well as regional visioning. The key challenge is to adequately identify accessibility not just as a single performance measure among dozens but to elevate it to a primary goal in its own right. And ideally, the subservient nature of mobility as a transportation planning goal would be recognized as well, with mobility as a goal only to the extent that it contributes to accessibility improvements.

These principles were partially followed in the Virginia Department of Transportation's SMART Scale (System Management and Allocation of Resources for Transportation) project prioritization process.[36] The SMART Scale program began when political leaders in Virginia sought to invest their transportation infrastructure dollars more judiciously through a transparent and objective process. The Virginia Legislature passed House Bill 2, which mandated an objective and quantifiable process for the allocation of transportation-infrastructure dollars. The specifics of the policies to implement House Bill 2 were developed by the Commonwealth Transportation Board in 2015.

The goals of the board were for the state to fully fund high-scoring projects proposed by local governments and metropolitan planning agencies. It sought project-prioritization criteria that would link transportation and land-use decision making, that would be capable of evaluating projects of all modes, and that could be used to compare small and large projects alongside one another, so that competitiveness in the project-selection process would be unaffected by project size.

The board decided on six major dimensions for consideration: safety, congestion mitigation, accessibility, environmental quality, economic development, and land use.[37] (See table C.1.) To ensure simplicity, just one to three measures were used to score each of these dimensions. This parsimoniousness facilitated stakeholders' capacity to quickly visualize the project-prioritization system and to understand the relative importance of the various criteria. Different regions of the state were permitted to weight the value of the six dimensions differently.[38] Some areas placed more focus on economic development, while others placed more focus on congestion mitigation. The state's recommended weights placed more emphasis on congestion mitigation (45 percent of the score) than on accessibility (15 percent of the score) for major metropolitan areas. Thus, while accessibility was represented explicitly in decision making, mobility remained an independent goal with triple the weight of accessibility. Other goals such as safety or environmental quality represented distinct and important dimensions of the transportation system not captured by the accessibility goal.

The accessibility dimension of the project-selection criteria included three measures: accessibility to jobs, accessibility to jobs for disadvantaged persons, and access to multimodal choices. As with the Puget Sound plan, the latter measure, which focuses on access to the transportation system itself, captures people's ability to reach destinations poorly. Furthermore, a complete evaluation of the accessibility effects of transportation projects demands attention to their land-use impacts as well. This step was not present in the Virginia process.

TABLE C.1 Virginia Department of Transportation's SMART Scale Project prioritization criteria

FACTOR AREAS	MEASURE ID	MEASURES
Safety	S.1	Number of fatal and injury crashes (50%)
	S.2	Rate of fatal and injury crashes (50%)
Congestion mitigation	C.1	Person throughput (50%)
	C.2	Person hours of delay (50%)
Accessibility	A.1	Access to jobs (60%)
	A.2	Access to jobs for disadvantaged persons (20%)
	A.3	Access to multimodal choices (20%)
Environmental quality	E.1	Air quality and environmental effect (50%)
	E.2	Impact to natural and cultural resources (50%)
Economic development	ED.1	Project support for economic development (60%)
	ED.2	Intermodal access and efficiency (20%)
	ED.3	Travel time reliability (20%)
Land use[a]	L.1	Transportation-efficient land use (100%)

Source: Virginia Department of Transportation, *SMART Scale Technical Guide* (Richmond, VA: Virginia Department of Transportation, 2016).

[a] For areas over 200,000 in population

Furthermore, the weight accorded to accessibility in the ranking system was relatively low, at least relative to the weight accorded to the mobility component.

In the first round of project selection, in 2016, the SMART Scale process yielded 321 projects, out of which 163 were eventually funded. The selected projects included a range of types, including pedestrian, transportation-demand-management, and roadway projects. Both large and small projects were selected, and several of the funded projects came in at under $1 million.

The Virginia SMART Scale process was notable for its incorporation of measured accessibility as a primary criterion for project evaluation. The process would be more transformative still if congestion mitigation were not also a primary criterion, but rather was considered within the context of congestion's accessibility impacts. The formal weighting scheme suggests a route for incremental progress toward accessibility-based planning: gradually increasing the relative weighting of accessibility in comparison with mobility as a measure of transportation benefit.

Overcoming Obstacles to Reform

Yet, in practice, important constituencies guard mobility-based planning from significant reform. In many planning contexts in the United States and elsewhere, the most powerful of these are local constituencies, such as current residents, neighborhood organizations, and business owners, who mobilize to pressure public officials to enact and maintain regulations limiting development density, in part to keep local congestion at bay. Restrictions on development densities inexorably increase the spread of metropolitan regions, with attendant deleterious effects on accessibility, even by car (see chapter 4). These constituencies' quest to protect their localized mobility ends up degrading the performance of the transportation system when viewed from a broader perspective.

These constituencies are nevertheless rational in seeking to protect their own mobility. New development can pit the interests of incumbent residents against those of would-be newcomers or against the residents of the region as a whole. Consider, for example, a proposed higher-density residential development in an established neighborhood near downtown or a major public-transport station. The development might be analyzed as highly beneficial in accessibility terms, much more so than if it were located at other potential sites in the region. Yet, from the standpoint of incumbent residents of the neighborhood, there is little accessibility to be gained and much to be lost. Since the development is residential, it adds little to the destinations reachable by existing residents, but the new residents may add to roadway congestion, thereby degrading the accessibility of the incumbents. Neighborhood resistance to development is thus readily explicable in accessibility terms: although accessibility gains will accrue to the newcomers and to the region as a whole, existing residents may only see losses.

Though the local exclusionary impulse is not a direct product of planning institutions, it is at least in part shaped by them. Formalized mobility planning legitimates and gives a methodological language—together with a professional cadre to speak it—to the desire of congestion-mobilized neighbors for exclusion. Yet the impulse of each neighborhood to protect its accessibility by excluding traffic-generating land uses restricts the development of locations with the highest accessibility and therefore results in degraded regional accessibility. In this sense, this transportation/land-use conundrum amounts to a tragedy of the commons, where the actions of individuals who rationally seek their own self-interest undermine a commonly shared good or resource, a core rationale for planning to begin with.[39] While acknowledging local interests, planners and policy makers can aspire to institutions that facilitate greater regional (and global) representation.

Professional insistence on the priority of accessibility in transportation and land-use planning can be one of those institutions. When accessibility is the rubric guiding planning decisions, the traffic performance of a set of nearby intersections is subordinated to the broader interest of residents regionwide in reaching destinations. Where incumbent residents will still be concerned about roadway delay, that local focus will be counterbalanced rather than reinforced and legitimated by the language, models, and regionally scaled metrics of planners.[40]

Reform of professional institutions does not come easily. Professions can retain a vested interest in the status quo that has often served them well, generating both the need for their services and the flow of resources to pay for them. Yet the examples of accessibility deployment described in this chapter provide evidence of a changing professional culture, as did the appearance of accessibility (albeit as a dimension of mobility) in the venerable *Highway Capacity Manual*, published in 2016.[41] A demonstration of the potential for a thoroughgoing professional transformation in a transportation organization had occurred two decades earlier when the Institute of Transportation Engineers, which had previously been oriented toward rapid roadway travel, gave its imprimatur to traffic calming, a set of techniques designed specifically to slow cars down.[42] Professionals in practice frequently need the legitimacy offered by their field's organizations before they can alter their practices; professional organizations in transportation and land-use fields are thus important actors in the accessibility shift, which can be assisted by three principal approaches.

Bottom-Up

Planners in local and regional practice can inject accessibility thinking, metrics, and models into day-to-day planning operations. There is some evidence that this is already occurring.[43] This bottom-up shift is more than simple planning for multimodalism or mixed-use development, as welcome as those innovations may be. Rather, it is a shifting of the rationale for what constitutes success in transportation. Cases such as the San Francisco Bicycle Plan are laudable but fail to challenge the evaluative mobility framework, which impedes new solutions. By contrast, a shift to accessibility-based thinking could enable a host of innovations in transportation and land-use practice by redefining success.

Top-Down

The planner who seeks to analyze decisions in accessibility terms is in effect volunteering for extra duty: the norms of mobility-based analysis remain in place and, because of the lack of mandates, few resources are devoted to accessibility-based analysis. Thus, observed instances of bottom-up accessibility-based reform, while important, need bolstering of a different sort. "Top-down" reform refers to altering the mandates and norms promulgated by state and national agencies, boards of directors, funding organizations, and professional societies.

This can generate the requirements for accessibility-based planning, provide resources, and, importantly, offer professional legitimacy to the would-be accessibility planner.

Outside-In

Sectors including real estate and economic development are keenly aware of the role of accessibility in shaping outcomes of concern to them. The medical and public health fields are invested in the ability of patients to reach medical care. Outside-in reform is based on bridge-building to realms such as these outside of traditional transportation and land-use planning. The interests of these sectors naturally align with people's need to reach valued destinations. For this reason, they can provide needed support for planning reform.

Reform of analytic frameworks, via any of these routes, can shift outcomes in at least three ways. First, it can aid in "making the good easy"; that is, it can remove obstacles to accessibility-increasing transportation investments and land-use reforms.[44] Mobility-based evaluation may not foreclose these options in any absolute sense, but it impedes accessibility-increasing policies by imposing the burden of proving that their mobility impacts will not be too onerous. Shifting to accessibility-based evaluation allows these policies to be judged on their accessibility merits. Second, in the realms of transportation and land use, accessibility can serve as a coordinating principle for policy across multiple agencies and governmental levels. It can help ensure that policy at various levels does not work at cross purposes to produce perverse transportation outcomes. Third, altering the criteria can help shift the broader discourse around transportation and land use. When a dense development in a high-accessibility location is seen not as degrading LOS but rather as improving accessibility, public discourse about what success means in transportation planning is transformed. Such language will not be adopted immediately or universally, and it will not resolve questions pertaining to accessibility for whom, where, and by which modes. But such language can, over time, shift the thinking of some planning-process participants and stakeholders, just as the technical language of mobility has shaped discourse for decades.

Accessibility and Planning Goals

Mobility language has served planning poorly. Ultimately, planners seek transportation and land-use systems to support thriving cities and regions along multiple dimensions, including environmental protection, social equity, and economic development—or the "planner's triangle."[45] Quite apart from its mismatch with the purpose of transportation, mobility as a guiding principle for transportation and land-use planning is mismatched with all three dimensions. By elevating movement—the inherently energy-intensive means of providing accessibility—to an end, it sidelines the potential roles of proximity and connectivity in furthering environmental goals. Since mobility is an attribute of transportation links rather than people, it masks the distributional consequences of transportation decisions and thus impedes social-equity planning. Economic development is sometimes seen as mobility planning's forte, but the driving force behind agglomeration economies, the economic engine of cities and regions, is not movement per se, but the potential for interaction, or accessibility.[46]

As an alternative to transportation planning's mobility logic, VKT reduction fits well with environmental-protection goals but has little to say about either social equity or economic development. By contrast, the accessibility shift fits all three corners of the planner's triangle. Accessibility-based planning holds the potential to reduce the environmental impact of the transportation system by prioritizing proximity and connectivity, which, compared to mobility, are the less energy-intensive means of supporting interaction. But in contrast with VKT reduction, accessibility promotion does not limit itself to the austere value of travel reductions alone. Instead, it aims to augment the capacity of the built environment to offer a high quality of life by supporting people's interaction with a range of destinations and the activities available at them. Because accessibility is an attribute of people and neighborhoods, not of intersections and freeway links, it is always distributed unevenly between socioeconomic groups and geographic regions and thus lends itself to equity-based evaluations. The evaluation of accessibility also inherently incorporates dimensions of the urban economy, since increasing the potential for interaction is the central function of cities to begin with. Since knowledge spillovers and benefits to agglomeration—both products of this interaction—are principal drivers in economic development, accessibility also supports the planner's economic-development goals.

In none of these realms will the accessibility shift do the job on its own. To have an impact, changing definitions of transportation and land-use success need to facilitate reforms in the built environment and the policies that operate within it. The accessibility shift cannot, in and of itself, provide guidance on the relative importance to be accorded to accessibility by the various modes; at different locales; or of people of differing incomes, classes, or races. It cannot resolve resource conflicts between transportation and other social goods such as education, health, or public safety. It amounts to the simple proposition that when planning aims to gauge and value the benefits of transportation, it will do so in accessibility rather than in mobility terms. While such a shift cannot provide the answers for planners and policy makers in their search for better cities and regions, in the realm of transportation and land-use planning, it can, at long last, help them ask the right questions.

Appendix A

PROCEDURE FOR ACCESSIBILITY ANALYSIS FOR LAND-USE PROJECTS

Step 1. Calculate Total Accessibility for the Region

This method employs a gravity-based accessibility formula to measure total regional accessibility to all employment (see equation A.1). A more complete description of this method is available in *Transport Policy*.[1] This measure is widely used in the research literature and is relatively straightforward to compute when interzonal travel-time data are available. The Southeast Michigan Council of Governments provided population and employment counts by traffic-analysis zone (TAZ) and auto travel times between zones by time of day: a.m. peak, p.m. peak, and off peak. Based on research that suggests a relationship between metropolitan population size and work travel impedance,[2] a range of impedance coefficients typical for work travel for smaller metropolitan areas was used: 0.10, 0.13, and 0.16, with 0.13 as the default value.

The accessibility of each zone is a function of the number of destinations in other zones and the impedance of travel between zones (see equation A.1). Total regional accessibility is the sum of accessibility from each zone multiplied by the resident population in each zone. Total accessibility, rather than average accessibility, is used in this analysis because the effects of population growth are incorporated at later stages through the elasticity calculation. Total accessibility necessarily increases with any zonal increase in population or employment, assuming that travel impedances are held constant.

$$total\ accessibility\ before = \sum_i p_i a_i = \sum_i p_i \sum_i d_j e^{-0.13t_{ij}} \quad (A.1)$$

Summing over all zones i, p_i is the total residential population in each zone i, and a_i is the accessibility reachable from each zone i. The second part of the equation further breaks down the accessibility component for each zone i, a_i. Each a_i is the sum over zones j of the number of destinations in each zone j, d_j, times the impedance between zones i and j, which here has the negative exponential form of the travel time from zone i to zone j (t_{ij}).

Step 2. Conduct a Traffic-Impact Analysis for the Development

Local jurisdictions in the United States typically require a traffic-impact analysis for any sizable development. These analyses forecast additional delay expected to result from the development. In urban areas, traffic-impact analyses result in reports of additional seconds of delay for each type of movement through a set of affected intersections.[3] Delays are aggregated up to the intersection level by weighting each type of intersection movement with the projected number of trips making that movement, resulting in an average delay for the entire intersection. Any zone-to-zone route that is expected to traverse this particular intersection is assigned this additional aggregated delay, as described in the next step.

Step 3. Update Regional Travel Pattern Data Based on Information from the Traffic-Impact Analysis

Intersection-level delays from the previous step are assigned to zone-to-zone travel times via a routing procedure.[4] This is completed by calculating the route from every TAZ to every other TAZ and identifying which routes run through the affected intersections. If a TAZ-to-TAZ route runs through an affected intersection, its travel time is updated to include the intersection-level delay. In addition, if the shortest route no longer runs through an affected intersection, the routing procedure identifies the updated fastest route. This computationally intensive procedure is implemented with programming and makes use of open-source mapping and routing data. The result is a set of new zone-to-zone travel times taking into account the delays added by the proposed development.

Step 4. Change Regional Land-Use Data Based on Information from the Development

Given the proposed development's location and size, either population or employment can be added to the appropriate zone. Translation of housing units into a population change is accomplished on the basis of census data on average household size for each particular type of housing unit within the development. To convert retail development size into employment totals, the *Planners Estimating Guide* suggests that each retail job requires approximately 47 square meters (510 square feet) of space.[5]

With new forecast information on land-use and transportation patterns that will result after the project, a new, "after" total regional accessibility can be calculated. The same gravity accessibility formula is used, but with new travel times, population, and employment (as destination activities) information (see equation A.2).

In most cases, the total accessibility after will be higher, because additional population (or employment) has been added. However, an increase in total accessibility does not necessarily imply an improvement in average accessibility. The accessibility gain must be compared with the size of the development to examine its effect on the average. This is the purpose of the next step, calculating accessibility elasticity.

$$\textit{total accessibility after} = \sum_i (p_i + \Delta p_i) \sum_i e^{-0.13(t_{ij} + \Delta t_{ij})} (d_j + \Delta d_j) \qquad \text{(A.2)}$$

Total accessibility after is the sum over all zones i of the new population in each zone i ($p_i + \Delta p_i$) times the accessibility available from each zone i. The new accessibility for each zone i is the sum of the new destination totals ($d_j + \Delta d_j$) times the new travel time impedances $e^{-0.13(t_{ij} + \Delta t_{ij})}$.

Step 5. Calculate the Accessibility Elasticity

Accessibility elasticity is computed as the proportional change in accessibility divided by the proportional change in regional size. Regional size is measured either by population or employment. The formula for accessibility elasticity appears in equation A.3. The accessibility elasticity score minus 1 yields the percentage by which the development outperforms or underperforms the current average accessibility of the region.

$$\textit{accessibility elasticity} = \frac{\left(access^{after} - access^{before}\right) / \, access^{before}}{\left(population^{after} - population^{before}\right) / \, population^{before}} \qquad \text{(A.3)}$$

TABLE A.4A. Metropolitan regions included in the study

REGION	POPULATION RANK	2009 MSA POPULATION
New York	1	19,069,796
Los Angeles	2	12,874,797
Chicago	3	9,580,567
Dallas	4	6,447,615
Philadelphia	5	5,968,252
Houston	6	5,867,489
Washington, DC	8	5,476,241
Atlanta	9	5,475,213
Boston	10	4,588,680
Detroit	11	4,403,437
Phoenix	12	4,364,094
San Francisco	13	4,317,853
Seattle	15	3,407,848
Minneapolis	16	3,269,814
San Diego	17	3,053,793
Baltimore	20	2,690,886
Denver	21	2,552,195
Portland	23	2,241,841
Cincinnati	24	2,171,896
Cleveland	26	2,091,286
Orlando	27	2,082,421
San Antonio	28	2,072,128
Kansas City	29	2,067,585
Las Vegas	30	1,902,834
Columbus	32	1,801,848
Charlotte	33	1,745,524
Indianapolis	34	1,743,658
Virginia Beach	36	1,674,498
Nashville	38	1,582,264
Memphis	41	1,304,926
Louisville	42	1,258,577
Richmond	43	1,238,187
Oklahoma City	44	1,227,278
Hartford	45	1,195,998
New Orleans	46	1,189,981
Buffalo	50	1,123,804
Rochester	51	1,035,566
Tucson	52	1,020,200

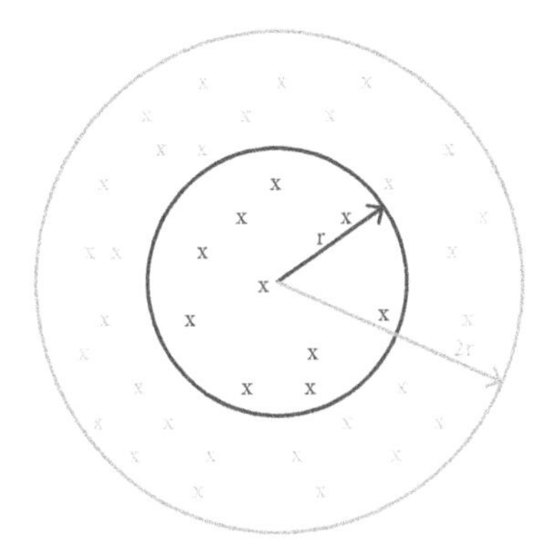

	Radius	Area	Density	Destinations or accessibility
Smaller circle	r	πr^2	$10/\pi r^2$	10
Larger circle	2r	$4\pi r^2$	$10/\pi r^2$	40

FIGURE A.4A. Illustration of speed effect of accessibility (holding destinations constant). One "x" represents one destination.

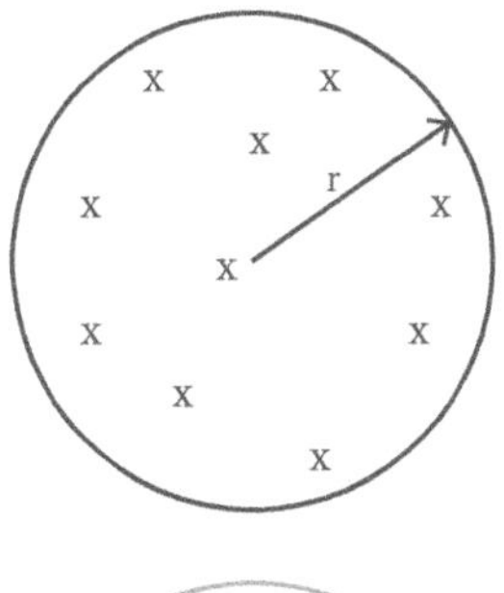

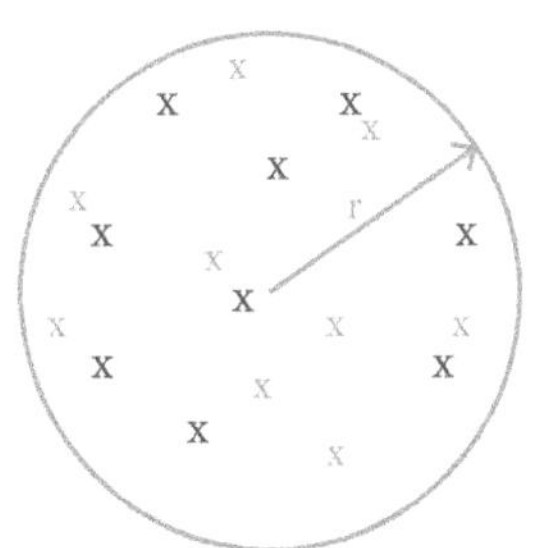

	Radius	Area	Density	Destinations or accessibility
Top circle	r	πr^2	$10/\pi r^2$	10
Bottom circle	r	πr^2	$20/\pi r^2$	20

FIGURE A.4B. Illustration of density effect of accessibility (holding speeds constant). One "x" represents one destination.

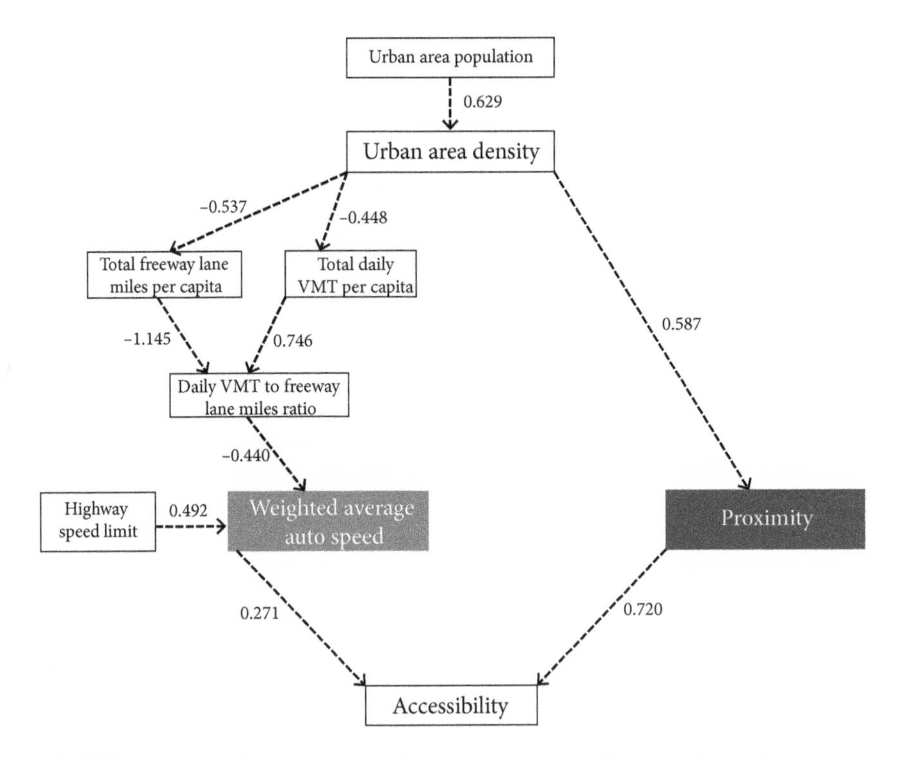

FIGURE A.4C. Path analysis of relationships of density, speed, proximity, and accessibility

Variables in figure A.4c are defined as follows:

Urban area density: Population of the urbanized area divided by its total land area in square kilometers, retrieved from *Highway Statistics 2008* by the Federal Highway Administration.[6]

Proximity: Median gravity-based work accessibility index when distance between origins and workplace destinations (rather than travel time) is used as the metric of impedance.

Accessibility: Median accessibility score for residents in the region, calculated as a gravity model with peak-period travel time by automobile between origins and workplace destinations used as the metric of impedance.

Highway speed limit: The speed limit of the state or territory to which each region belongs. Among the 38 MPOs in the study, this variable takes one of three values: 65 mph, 70 mph, or 75 mph.

Total daily VMT per capita: Total daily vehicle-miles traveled by the residents of the region divided by the total population of the region, retrieved from *Highway Statistics 2008* by the Federal Highway Administration.

Total freeway lane miles per capita: Total freeway lane miles within the region divided by the total population of the region, retrieved from *Highway Statistics 2008* by the Federal Highway Administration.

Total daily VMT to total lane miles ratio: Total daily VMT per capita divided by total freeway lane miles per capita.

Weighted average auto speed: The average speed, weighted by the imputed travel volume share for the corresponding zonal pair.[7]

The values represented along each link in the figure are standardized regression coefficients, a measure of the strength of the relationship between the variables shown in the diagram. The dependent variable of each regression equation is the variable to which an arrow points. Independent variables are those represented as pointing toward the dependent variable. For example, "weighted average auto speed" is the dependent variable in a regression with "highway speed limit" and "daily VMT to total lane miles ratio" as the independent variables.

In a path analysis (appendix A, fig. 4.c), weights along sequential paths are multiplied to calculate the weight (or strength of relationship) along the entire link; weights of parallel paths are summed. Thus, the weight from density to speed is calculated as:

$$[(-0.537*-1.145) + (-0.448*0.746)]*(-0.440) = -0.123$$

As predicted, this link is weak relative to the other links shown in the figure, because it incorporates two countervailing factors: low-density regions are freeway-rich on a per-capita basis, but these regions simultaneously demonstrate high VMT per capita.

The net effect of density on accessibility is the sum of the positive effect via higher proximity and the negative effect via slower speeds. The effect of density on accessibility can be analyzed by comparing the composite weight along the right-hand path (via proximity) and the left-hand path (via speed). The weight along the entire speed path equals

$$(-0.123*0.271) = -0.033,$$

while that along the proximity path equals

$$(0.587*0.720) = 0.423.$$

TABLE A.5A. Metropolitan area summary data, in rank order of public-transport accessibility (2014)

METRO AREA	TIME-WEIGHTED JOB ACCESSIBILITY VIA TRANSIT	85TH-PERCENTILE 30-MINUTE JOB ACCESSIBILITY VIA TRANSIT	POPULATION (MILLIONS)	RAIL VRM PER CAPITA	BUS VRM PER CAPITA	PERCENT IN HIGH DENSITY	FIT INDEX	GINI INDEX
New York	47,536	210,186	19.0	27.9	16.9	31.8%	0.34	0.81
San Francisco	14,400	65,246	4.4	19.1	15.0	5.6%	0.22	0.76
Los Angeles	11,811	43,430	13.0	2.5	18.1	3.8%	0.15	0.67
Washington	11,505	47,759	5.6	14.5	18.3	2.5%	0.17	0.83
Chicago	11,418	48,116	9.4	12.3	13.4	3.3%	0.17	0.80
Boston	11,015	49,237	4.5	11.7	11.8	4.4%	0.30	0.81
Philadelphia	7,845	35,217	6.0	7.7	12.8	4.0%	0.14	0.80
Seattle	6,598	26,141	3.5	3.0	27.4	0.8%	0.28	0.81
Denver	5,466	20,467	2.6	4.1	16.3	0.1%	0.13	0.71
San Jose	5,076	16,441	1.9	2.9	11.0	0.4%	0.13	0.51
Portland	4,858	18,723	2.3	3.8	14.3	0.1%	0.22	0.70
Milwaukee	4,735	19,216	1.6	0.0	14.3	0.2%	0.36	0.67
Minneapolis	4,476	17,651	3.3	1.1	14.8	0.2%	0.36	0.78
Baltimore	4,226	14,633	2.8	3.7	13.5	1.0%	0.14	0.82
Miami	4,000	15,333	5.6	2.5	14.7	1.7%	0.07	0.67
Houston	3,940	15,166	5.9	0.2	12.1	0.4%	0.04	0.81
Salt Lake City	3,824	13,733	1.1	7.4	11.3	0.0%	0.21	0.61
San Diego	3,536	12,752	3.0	3.4	14.6	0.8%	0.26	0.65
Phoenix	2,936	9,981	4.3	0.6	9.7	0.1%	0.08	0.70
Sacramento	2,898	11,052	2.2	1.8	7.2	0.0%	0.15	0.73
Dallas	2,846	10,113	6.5	1.7	8.3	0.5%	0.02	0.76
Pittsburgh	2,779	12,268	2.4	2.0	13.3	0.0%	0.15	0.84
San Antonio	2,650	9,849	2.2	0.0	15.0	0.0%	0.15	0.74
Las Vegas	2,626	8,162	2.0	1.4	12.7	0.4%	0.06	0.66

Buffalo	2,486	10,432	1.1	0.9	9.0	0.0%	0.28	0.67
Cleveland	2,266	8,372	2.1	1.9	10.7	0.2%	0.14	0.71
Columbus	2,233	9,506	1.9	0.0	8.5	0.2%	0.12	0.77
St. Louis	2,184	7,804	2.8	2.2	10.4	0.0%	0.14	0.76
Hartford	2,137	9,768	1.2	0.0	9.5	0.3%	0.29	0.77
Atlanta	1,910	6,995	5.4	3.3	8.6	0.0%	0.08	0.87
Providence	1,871	8,252	1.7	0.0	9.6	0.4%	0.22	0.69
New Orleans	1,844	8,364	1.2	0.8	6.7	0.2%	0.13	0.71
Tampa	1,746	6,865	2.8	0.0	8.7	0.1%	0.04	0.69
Detroit	1,736	6,373	4.3	0.1	7.1	0.0%	0.03	0.62
Charlotte	1,672	6,541	1.8	0.5	11.9	0.0%	0.09	0.84
Louisville	1,639	6,509	1.3	0.0	10.0	0.0%	0.22	0.78
Kansas City	1,624	6,696	2.1	0.2	7.0	0.0%	0.15	0.74
Indianapolis	1,607	6,263	1.7	0.0	5.9	0.0%	0.09	0.76
Austin	1,590	7,001	1.8	0.2	11.0	0.4%	0.15	0.83
Raleigh	1,432	6,372	1.2	0.0	7.2	0.0%	0.02	0.83
Cincinnati	1,421	5,792	2.1	0.0	7.9	0.0%	0.21	0.79
Orlando	1,329	4,921	1.9	0.1	13.4	0.0%	0.02	0.74
Nashville	1,250	5,232	1.6	0.1	6.4	0.0%	0.08	0.85
Virginia Beach	1,200	4,784	1.7	0.2	9.3	0.0%	0.07	0.64
Riverside	1,105	4,346	4.1	0.5	6.2	0.0%	0.25	0.61
Birmingham	705	2,828	1.1	0.0	4.1	0.0%	0.03	0.82

Sources: Andrew Owen and David M. Levinson, *Access Across America: Transit 2014 Data* (Minneapolis: University of Minnesota Digital Conservancy, 2014), retrieved from http://dx.doi.org/10.13020/D6MW2Q; Federal Transit Administration, *National Transit Database* (Washington, DC, 2014), retrieved from https://www.transit.dot.gov/ntd; Minnesota Population Center, *National Historical Geographic Information System* (2011).

Note: Rail is defined as synonymous with the NTD's fixed-guideway service, and bus is defined as the NTD's non-fixed-guideway service.

TABLE A.7A. Correspondence between NHTS travel purposes and NAICS job classifications

NHTS TRIP PURPOSES/NAICS JOB CLASSIFICATION	PERCENT OF NONWORK TRIPS
Buy goods: groceries/clothing/hardware store; shopping	28.8
New car dealers; used car dealers, motorcycle, ATV, and personal watercraft dealers; boat dealers; all other motor vehicle dealers; automotive parts and accessories stores; tire dealers; furniture stores; floor covering stores; window treatment stores; all other home furnishings stores; household appliance stores; radio, television, and other electronics stores; computer and software stores; camera and photographic supplies stores; home centers; paint and wallpaper stores; hardware stores; other building material dealers; outdoor power equipment stores; nursery, garden center, and farm supply stores; supermarkets and other grocery (except convenience) stores; convenience stores; meat markets; fish and seafood markets; fruit and vegetable markets; baked goods stores; confectionery and nut stores; all other specialty food stores; beer, wine, and liquor stores; pharmacies and drugstores; cosmetics, beauty supplies, and perfume stores; optical goods stores; food (health) supplement stores; all other health and personal care stores; men's clothing stores; women's clothing stores; children's and infants' clothing stores; family clothing stores; clothing accessories stores; other clothing stores; shoe stores; jewelry stores; luggage and leather goods stores; sporting goods stores; hobby, toy, and game stores; sewing, needlework, and piece goods stores; musical instrument and supplies stores; book stores, news dealers and newsstands; prerecorded tape, compact disc, and record stores; department stores (except discount department stores); discount department stores; warehouse clubs and supercenters; all other general merchandise stores; florists, office supplies and stationery stores; gift, novelty, and souvenir stores; used merchandise stores; art dealers; manufactured (mobile) home dealers; tobacco stores; all other miscellaneous store retailers (except tobacco stores)	
Get/eat meal; meals; social event	12.9
Full-service restaurants; limited-service restaurants; cafeterias, grill buffets, and buffets; mobile food services	
Visit friends/relatives	11.3
Population (rather than NAICS codes) is considered as the number of destinations in this category.	

NHTS TRIP PURPOSES/NAICS JOB CLASSIFICATION	PERCENT OF NONWORK TRIPS
Go to gym/exercise/play sports	7.1
Golf courses and country clubs; skiing facilities; fitness and recreational sports centers; bowling centers	
Buy services: video rentals/dry cleaner/post office/car service/bank	7.0
Postal service; couriers and express delivery services; commercial banking; savings institutions; credit unions; other depository credit intermediation; consumer lending; mortgage and nonmortgage loan brokers; consumer electronics and appliances rental; formal wear and costume rental; video tape and disc rental; general rental centers; general automotive repair; automotive exhaust system repair; automotive transmission repair; other automotive mechanical and electrical repair and maintenance; automotive body, paint, and interior repair and maintenance; automotive glass replacement shops; automotive oil change and lubrication shops; car washes; all other automotive repair and maintenance; consumer electronics repair and maintenance; computer and office machine repair and maintenance; communication equipment repair and maintenance; other electronic and precision equipment repair and maintenance; home and garden equipment repair and maintenance; appliance repair and maintenance; reupholstery and furniture repair; footwear and leather goods repair; other personal and household goods repair and maintenance; coin-operated laundries and drycleaners; drycleaning and laundry services (except coin-operated); pet care (except veterinary) services; photofinishing laboratories (except one-hour); one-hour photofinishing	
Go to school as student	6.5
Elementary and secondary schools; junior colleges, colleges, universities, and professional schools; business and secretarial schools; computer training; professional and management development training; cosmetology and barber schools; flight training; apprenticeship training; other technical and trade schools; fine arts schools; sports and recreation instruction; language schools; exam preparation and tutoring; automobile driving schools; all other miscellaneous schools and instruction; educational support services	
Go to religious activity and school/religious activity	4.9
Funeral homes and funeral services; cemeteries and crematories; religious organizations	

(Continued)

TABLE A.7A. Continued

NHTS TRIP PURPOSES/NAICS JOB CLASSIFICATION	PERCENT OF NONWORK TRIPS
Go out/hang out: entertainment/theater/sports event/bar	4.3
Motion picture theaters (except drive-ins); drive-in motion picture theaters; theater companies and dinner theaters; dance companies, musical groups and artists; other performing arts companies; sports teams and clubs; racetracks; other spectator sports; amusement and theme parks; amusement arcades; casinos (except casino hotels); other gambling industries; marinas; all other amusement and recreation industries; drinking places (alcoholic beverages)	
Buy gas	3.4
Gasoline stations with convenience stores; other gasoline stations	
Family personal business/obligations	3.2
Offices of real estate agents and brokers; offices of real estate appraisers; other activities related to real estate; passenger car rental; offices of notaries; title abstract and settlement offices; architectural services; landscape architectural services; building inspection services; interior design services; photography studios, portrait; employment placement agencies; temporary help services; travel agencies; investigation services; solid waste landfill; labor unions and similar labor organizations; political organizations; other similar organizations (except business, professional, labor, and political organizations); courts; police protection; legal counsel and prosecution; correctional institutions; parole offices and probation offices	
Medical/dental services	2.9
Offices of physicians (except mental health specialists); mental health specialists; offices of dentists; offices of chiropractors; offices of optometrists; offices of mental health practitioners (except physicians); offices of physical, occupational and speech therapists, and audiologists; offices of podiatrists; offices of all other miscellaneous health practitioners; family planning centers; outpatient mental health and substance abuse centers; HMO medical centers; kidney dialysis centers; freestanding ambulatory surgical and emergency centers; all other outpatient care centers; medical laboratories; diagnostic imaging centers; home health care services; ambulance services; blood and organ banks; all other miscellaneous ambulatory health care services; general medical and surgical hospitals; psychiatric and substance abuse hospitals; specialty (except psychiatric and substance abuse) hospitals; nursing care facilities; residential mental retardation facilities; residential mental health and substance abuse facilities; continuing care retirement communities; homes for the elderly; other residential care facilities; services for the elderly and persons with disabilities; other individual and family services; emergency and other relief services; vocational rehabilitation services	

NHTS TRIP PURPOSES/NAICS JOB CLASSIFICATION	PERCENT OF NONWORK TRIPS
Coffee/ice cream/snacks	1.4
Snack and nonalcoholic beverage bars	
Visit public place: historical site/museum/park/library; go to library: school related	1.3
Libraries and archives; museums; historical sites; zoos and botanical gardens; nature parks and other similar institutions	
Day care	1.0
Child day care services	
Rest or relaxation/vacation	1.0
Scheduled passenger air transportation; nonscheduled chartered passenger air transportation; deep-sea passenger transportation; coastal and Great Lakes passenger transportation; inland water passenger transportation; scenic and sightseeing transportation, land; scenic and sightseeing transportation, water; scenic and sightseeing transportation, other; tour operators; convention and visitors bureaus; hotels (except casino hotels) and motels; casino hotels; bed-and-breakfast inns; all other traveler accommodations; RV (recreational vehicle) parks and campgrounds; recreational and vacation camps (except campgrounds)	
Attend meeting: PTA/home owners association/local government	1.0
Civic and social organizations; legislative bodies; executive and legislative offices, combined; American Indian and Alaska native tribal governments; other general government support; other justice, public order, and safety activities; administration of public health programs; administration of human resources programs (except education, public health, and Veterans Affairs programs); Administration of Veterans Affairs; Administration of Housing programs; Administration of Urban Planning and Community and Rural Development; Administration of General Economic Programs; regulation, licensing, and inspection of miscellaneous commercial sectors	
Pet care: walk the dog/vet visits	0.9
Pet and pet supplies stores; veterinary services	
Use personal services: grooming/haircut/nails	0.5
Barber shops, beauty salons, nail salons, diet and weight-reducing centers, other personal care services	
Use professional services: attorney/accountant	0.5
Offices of lawyers, all other legal services, offices of certified public accountants, tax preparation services, other accounting services	

Appendix B

VARIABLES USED IN THE TEXT

VARIABLE NAME OR SYMBOL	DESCRIPTION
N_m	Estimated number of nonwork trips for metropolitan region m
85th-percentile 30-minute job accessibility via transit	The average number of jobs reachable within 30 minutes of travel time by public transport at the 85th percentile of the regional accessibility distribution
a^a_i	Accessibility index for travel by automobile for zone i
a^t_i	Accessibility index for travel by public transport for zone i
Accessibility	Median accessibility score for residents in the region, calculated as a gravity model with peak-period travel time by automobile between origins and workplace destinations used as the metric of impedance
a_i	Accessibility of each zone i
A_{region}	Total regional population-weighted accessibility to jobs
d_j	Destinations in each zone j

(*Continued*)

VARIABLE NAME OR SYMBOL	DESCRIPTION
E_r	The share of all national nonwork trips for purpose r
$f(t_{ij})$	Impedance between zones i and j
f_{ij}	Flow of vehicles from zone i to zone j over the peak hour
Fit Index	A dimensionless index that measures the correlation between block-level 30-minute public-transport accessibility and block-level gross population density
Gini Index	A dimensionless index that measures the evenness of the distribution of public-transport accessibility across the regional population
g_j	Total number of jobs located in zone j
h_i	The number of households residing in zone i
Highway speed limit	The speed limit of the state or territory to which each region belongs
i	Zonal index, residential zones
j	Zonal index, destination zones
J_{ir}	The number of jobs in zone i associated with trip purpose r
J_m	Number of jobs in metropolitan region m
MLMR	Mode-location match ratio
N_m	Estimated number of nonwork trips for metropolitan region m
N_r	The number of nonwork trips in a metropolitan region for trip purpose r
Percent in high density	Share of the regional population living in densities higher than 60 persons per gross acre
p_i	Total residential population in each zone i
Population	Total regional population
Proximity	Median gravity-based work accessibility index when distance between origins and workplace destinations (rather than travel time) is used as the metric of impedance
Rail VRM per capita	Rail service deployed, in per-capita vehicle-revenue miles (VRM) for rail modes
s^a_i	Share of the region's automobile-owning households located in zone i
s^t_i	Share of the region's transit-dependent households located in zone i

VARIABLE NAME OR SYMBOL	DESCRIPTION
T	Aggregate regional sum of peak-hour travel times
t_{ij}	Travel cost/time from zone i to j
Time-weighted job accessibility via transit	A weighted-average accessibility score for each metropolitan area based on the average number of jobs that are reachable by public transport within 10, 20, 30, 40, 50, and 60 minutes
Total daily VMT per capita	Total daily vehicle-miles traveled by the residents of the region divided by the total population of the region
Urban area density	Population of the urbanized area divided by its total land area in square kilometers
Weighted average auto speed	The average speed, weighted by the imputed travel volume share for the corresponding zonal pair
W_i	The nonwork attractiveness index at zone i

Notes

INTRODUCTION

1. Noah Kazis, a former writer for Streetsblog, offers a particularly apt example to illustrate the same point: "Imagine two drivers leaving downtown to head home. Each of them sits in traffic for the first ten miles of the commute but at that point, their paths diverge. The first one has reached home. The second has another twenty miles to drive, though luckily for her, the roads are clear and congestion doesn't slow her down. Who's got a better commute?" Most metrics of mobility—average speeds, percent of congested travel, peak-to-off-peak travel-time ratio, buffer index (the percent of travel time needed to ensure an on-time arrival)—would (perversely) assess the second driver to have the superior commute. Noah Kazis, "Report: Want to Ease Commuter Pain? Highways and Sprawl Won't Help," *Streetsblog NYC*, September 29, 2010, http://nyc.streetsblog.org/2010/09/29/report-want-to-ease-commuter-pain-highways-and-sprawl-wont-help/.

2. In other contexts, "accessibility" focuses on accommodating the needs of people with a range of abilities and is focused on building and facility design. By contrast, the use here pertains to the ability to reach destinations in a city or region and is focused on transportation and land use.

3. Reid Ewing," Measuring Transportation Performance," *Transportation Quarterly* 49, no. 1 (1995): 91–104.

4. David Schrank et al., *2015 Urban Mobility Scorecard* (College Station: Texas Transportation Institute, Texas A&M University, 2015).

5. Edward Weiner, *Urban Transportation Planning in the United States: A Historic Overview* (Westport, CT: Praeger, 1999).

6. Nathaniel Baum-Snow, "Did Highways Cause Suburbanization?," *Quarterly Journal of Economics* 122, no. 2 (2007): 775–805; Paul Waddell et al., "Modeling Interdependence in Household Residence and Workplace Choices," *Transportation Research Record: Journal of the Transportation Research Board*, no. 2003 (2007): 84–92; Transportation Research Board, *Expanding Metropolitan Highways: Implications for Air Quality and Energy Use*, Special Report 245 (Washington, DC: National Academies Press, 1995).

7. Michael D. Meyer and Eric J. Miller, *Urban Transportation Planning: A Decision-Oriented Approach*, 2nd ed. (New York: McGraw-Hill, 2001).

8. Peter R. Stopher and Arnim H. Meyburg, *Urban Transportation Modeling and Planning* (Lexington, MA: Lexington Books, 1975); Edward K. Morlok, *Introduction to Transportation Engineering and Planning* (New York: McGraw-Hill, 1978); Meyer and Miller, *Urban Transportation Planning*.

9. Robert B. Mitchell and Chester Rapkin, *Urban Traffic: A Function of Land Use* (New York: Columbia University Press, 1954).

10. Avijit Ghosh and C. Samuel Craig, "Formulating Retail Location Strategy in a Changing Environment," *Journal of Marketing* 47, no. 3 (1983): 53–68.

11. Hongbo Du and Corinne Mulley, "Relationship between Transport Accessibility and Land Value: Local Model Approach with Geographically Weighted Regression," *Transportation Research Record: Journal of the Transportation Research Board*, no. 1977 (2006): 197–205; Andy L. Krause and Christopher Bitter, "Spatial Econometrics, Land Values, and Sustainability: Trends in Real Estate Valuation Research," supplement, *Cities*

29, no. 2 (2012): S19–25; Paul Cheshire and Stephen Sheppard, "On the Price of Land and the Value of Amenities," *Economica* 62, no. 246 (1995): 247–67.

12. Jonathan Levine, *Zoned Out: Regulations, Market, and Choices in Transportation and Metropolitan Land Use* (Washington, DC: Resources for the Future, 2006).

13. For consistency with international terminology, we use metric units, except when quoting other documents or referring to data sources that employ Imperial units.

14. Ilan Salomon and Patricia L. Mokhtarian, "What Happens When Mobility-Inclined Market Segments Face Accessibility-Enhancing Policies?," *Transportation Research Part D* 3, no. 3 (1998): 129–40; Gerald L. Houseman, *The Right of Mobility* (Port Washington, NY: Kennikat Press, 1979), 19.

15. Susan L. Handy and Deb A. Niemeier, "Measuring Accessibility: An Exploration of Issues and Alternatives," *Environment and Planning A* 29, no. 7 (1997): 1176; Derek Halden, "Using Accessibility Measures to Integrate Land Use and Transport Policy in Edinburgh and the Lothians," *Transport Policy* 9, no. 4 (2002): 314.

16. Petter Naess, "'New Urbanism' or Metropolitan-Level Centralization?: A Comparison of the Influences of Metropolitan-Level and Neighborhood-Level Urban Form Characteristics on Travel Behavior," *Journal of Transport and Land Use* 4, no. 1 (2011): 25–44.

17. Martin Wachs and T. Gordon Kumagai, "Physical Accessibility as a Social Indicator," *Socio-Economic Planning Sciences* 7, no. 5 (1973): 439.

18. Jean-Paul Rodrigue, Claude Comtois, and Brian Slack. *The Geography of Transport System* (New York: Routledge, 2009), 54.

19. Irene Casas, "Evaluating the Importance of Accessibility to Congestion Response Using a GIS-Based Travel Simulator," *Journal of Geographical Systems* 5, no. 1 (2003): 110.

20. Transportation Research Board, *Highway Capacity Manual 2010*, http://hcm.trb.org/?qr=1.

21. WNYC Average Commute Times, from the 2006–10. U.S. Census Bureau's *American Community Survey* five-year estimates, https://project.wnyc.org/commute-times-us/embed.html.

22. Yingling Fan, "Household Structure and Gender Differences in Travel Time: Spouse/Partner Presence, Parenthood, and Breadwinner Status," *Transportation* 44, no. 2 (2017): 271–91.

23. Robert M. Haig, *Major Economic Factors in Metropolitan Growth and Its Arrangement: A Study of Trends and Tendencies in the Economic Activities within the Region of New York and Its Environs* (New York: Committee on Regional Plan of New York and Its Environs, 1927); Walter G. Hansen, "How Accessibility Shapes Land Use," *Journal of the American Planning Association* 25, no. 2 (1959): 73–76.

24. Geneviève Boisjoly and Ahmed M. El-Geneidy, "How to Get There?: A Critical Assessment of Accessibility Objectives and Indicators in Metropolitan Transportation Plans," *Transport Policy* 55 (2017): 38–50.

25. Karst T. Geurs and Bert van Wee, "Accessibility Evaluation of Land-Use and Transport Strategies: Review and Research Directions," *Journal of Transport Geography* 12, no. 2 (2004): 127–40.

26. Transportation Research Board, *Driving and the Built Environment: The Effects of Compact Development on Motorized Travel, Energy Use, and CO_2 Emissions* (Washington, DC: National Academies Press, 2009); Transportation Research Board, *Highway Capacity Manual 2010*, http://hcm.trb.org/?qr=1.

27. Institute of Transportation Engineers, *Transportation Impact Analyses for Site Development: An ITE Recommended Practice* (Washington, DC, 2010).

28. Gilles Duranton and Erick Guerra, *Developing a Common Narrative on Urban Accessibility: An Urban Planning Perspective* (Washington DC: Brookings, 2016).

29. Bent Flyvbjerg, *Making Social Science Matter: Why Social Inquiry Fails and How It Can Succeed Again* (New York: Cambridge University Press, 2001).

30. José Moore, Karen Lucas, and John Bates, *Social Disadvantage and Transport in the UK: A Trip-Based Approach* (Oxford: Transport Studies Unit, University of Oxford, 2013).

31. Bert van Wee and Karst T. Geurs, "The Role of Accessibility in Urban and Transport Planning," in *Handbook on Transport and Urban Planning in the Developed World*, ed. Michiel C. J. Bleimer, Corinne Mulley, and Claudine J. Moutou (Cheltenham, UK: Edward Elgar, 2016), 55–59.

32. While the accessibility shift pertains to all transportation modes, pedestrianism and cycling pose their own challenges for accessibility analysis, including the sparseness of available data and the prominence of difficult-to-measure safety and amenities in active transportation. These issues are beyond the scope of this book. Yet, even in the case of walking or cycling, there exist mobility and accessibility approaches to planning. For example, the *Highway Capacity Manual*'s assessments of level of service for nonmotorized facilities are based on pedestrian space (square feet of facility per pedestrian); bicycle-pedestrian meetings and passings per hour (a measure of potential interference on shared facilities); and, for bike facilities, meetings and active passings per minute, presence of a centerline, path width, and delayed passings. All these measure the quality of movement but shed no light on the ability of pedestrians and cyclists to reach destinations; an ample path that failed to offer any destinations within walking or cycling distance would still receive top scores. By contrast, shifting to accessibility in active transportation implies a move to a destination focus, which, for many walking and cycling facilities, would include recreation as a destination.

33. Derek Halden, "The Use and Abuse of Accessibility Measures in UK Passenger Transport Planning," *Research in Transportation Business & Management* 2 (2011): 12–19; Karel Martens, *Transport Justice: Designing Fair Transportation Systems* (New York: Routledge, 2017).

34. Charles E. Lindblom, "The Science of 'Muddling Through,'" *Public Administration Review* 19, no. 2 (1959): 79–88.

1. WHAT IS TRANSPORTATION FOR?

1. Jules Dupuit, "On the Measurement of the Utility of Public Works," in *Readings in Welfare Economics*, by Kenneth J. Arrow and Tibor Scitovsky (Homewood, IL: Richard D. Irwin, 1969), selected by a committee of the American Economic Association, translated in *International Economic Papers* 2 (1952): 83–110, English translation by R. H. Barback from "De la Mesure de l'Utilité des Travaux Publics," *Annales des Ponts et Chausées*, 2nd ser., vol. 8, 1844; Augustin Cournot, *Researches into the Mathematical Principles of the Theory of Wealth*, trans. Nathaniel T. Bacon with a Bibliography of Mathematical Economics by Irving Fisher (1838; repr. New York: MacMillan, 1897).

2. Alfred Marshall, *Principles of Economics* (New York: Macmillan, 1895).

3. Michael R. Bonavia, *The Economics of Transport* (London: Cambridge at the University Press, 1936), 4.

4. Paul A. Samuelson, "The Pure Theory of Public Expenditure," *Review of Economics and Statistics* 36, no. 4 (1954): 387–89.

5. Bonavia, *Economics of Transport*, 4.

6. Mitchell and Rapkin, *Urban Traffic*.

7. Mitchell and Rapkin's terms for the concept of derived demand in transportation were not as direct as those in use today. They clearly subordinated the trip to the desire to reach the destination. But like many seminal works viewed from the perspective of time, the book's treatment of its topic is more oblique than is generally remembered: "Psychological

analysis of travel motivations might include study of the formation of a tension within an individual (whether conscious or not) which he hopes to satisfy in the course of a trip, and the translation of this tension into an action decision. The decision process by which goal events and intended destinations are chosen should be explored. Attention should be given to the cognitive aspects of the potential destinations (locations or general localities where an individual is aware, believes or hopes that a goal event can occur) that are considered by the individual in choosing an intended destination." Mitchell and Rapkin, *Urban Traffic*, 54.

8. Mitchell, and Rapkin, *Urban Traffic*, 55.

9. Mitchell, and Rapkin, *Urban Traffic*, 105.

10. Patricia L. Mokhtarian and Ilan Salomon, "How Derived Is the Demand for Travel?: Some Conceptual and Measurement Considerations," *Transportation Research Part A* 35, no. 8 (2001): 695–719; Lothlorien S. Redmond and Patricia L. Mokhtarian, "The Positive Utility of the Commute: Modeling Ideal Commute Time and Relative Desired Commute Amount," *Transportation* 28, no. 2 (2001): 179–205; Salomon and Mokhtarian, "Mobility-Inclined Market."

11. Mokhtarian and Salomon, "How Derived Is the Demand for Travel?," 698.

12. Wachs and Kumagai, "Physical Accessibility as a Social Indicator," 439.

13. Randall Crane, "On Form versus Function: Will the New Urbanism Reduce Traffic, or Increase It?" *Journal of Planning Education and Research* 15, no. 2 (1996): 120.

14. Patricia L. Mokhtarian, Ilan Salomon, and Lothlorien S. Redmond, "Understanding the Demand for Travel: It's Not Purely 'Derived,'" *Innovation: The European Journal of Social Science Research* 14, no. 4 (2001): 355–80.

15. Salomon and Mokhtarian, "Mobility-Inclined Market Segments."

16. Mokhtarian and Salomon, "How Derived Is the Demand for Travel?," 699.

17. Patricia L. Mokhtarian and Ilan Salomon, "Travel for the Fun of It," *Access* 15 (1999): 27.

18. "The term [access] is not a good substitute for 'mobility', however, for it refers only to a limited number of circumstances in which access may be operative in place of physical mobility. . . . It is not even a good substitute for physical mobility, for it describes something other than mobility. In many cases, . . . it describes immobility. It can even mean, apparently without consultation with immobile persons, an assumption of the undesirability of a right of mobility for them. . . . The adoption of access as a social goal in place of mobility may at first blush appear to be a broader and more useful approach, but the most cursory examination shows that this is deceptive." Houseman, *Right of Mobility*, 20. "Most authors inside and outside transportation sciences see travel as an activity with a secondary utility, as opposed to those activities having primary utility. . . . To my thinking this is only partly true. Man is mobile. He cannot easily stay indoors all day long. He wants to 'exercise his legs', 'get a breath of fresh air' and feels satisfaction in the mere act of moving, in taking his body and mind from one place to another. We are, after all, descendants from the 'naked ape' who roamed the plains." Geurt Hupkes,"The Law of Constant Travel Time and Trip-Rates," *Futures*, 14, no. 1 (1982): 38–46.

19. Mokhtarian and Salomon, "How Derived Is the Demand for Travel?," 697.

20. Mokhtarian and Salomon, "How Derived Is the Demand for Travel?," 695–719.

21. Metropolitan Transportation Commission and Association of Bay Area Governments, *Bay Area Census*, http://www.bayareacensus.ca.gov/transportation/Commute time19802000.htm.

22. Levine, *Zoned Out*.

23. Michael Heilig et al., "Large-Scale Application of a Combined Destination and Mode Choice Model Estimated with Mixed Stated and Revealed Preference Data." *Transportation Research Record: Journal of the Transportation Research Board*, no. 2669 (2017):

31–40; Marcel Paulssen et al., "Values, Attitudes and Travel Behavior: A Hierarchical Latent Variable Mixed Logit Model of Travel Mode Choice," *Transportation* 41, no. 4 (2014): 873–88; David A. Hensher, "Measurement of the Valuation of Travel Time Savings," *Journal of Transport Economics and Policy* 35, no. 1 (2001): 71–98; Ron Dalumpines and Darren M. Scott, "Determinants of Route Choice Behavior: A Comparison of Shop versus Work Trips Using the Potential Path Area-Gateway (PPAG) Algorithm and Path-Size Logit," *Journal of Transport Geography* 59 (2017): 59–68; Brian H. Y. Lee et al., "Reexamining the Influence of Work and Nonwork Accessibility on Residential Location Choices with a Microanalytic Framework," *Environment and Planning A* 42, no. 4 (2010): 913–30; Barry Zondag and Marits Pieters, "Influence of Accessibility on Residential Location Choice," *Transportation Research Record: Journal of the Transportation Research Board*, no. 1902 (2005): 63–70; Patrick M. Schirmer, Michael A. B. van Eggermond, and Kay W. Axhausen, "The Role of Location in Residential Location Choice Models: A Review of Literature," *Journal of Transport and Land Use* 7, no. 2 (2014): 3–21; Issam Srour, Kara M. Kockelman, and Travis P. Dunn, "Accessibility Indices: Connection to Residential Land Prices and Location Choices," *Transportation Research Record: Journal of the Transportation Research Board*, no. 1805 (2002): 25–34; Joshua Auld and Abolfazl Mohammadian, "Planning-Constrained Destination Choice in Activity-Based Model: Agent-Based Dynamic Activity Planning and Travel Scheduling," *Transportation Research Record: Journal of the Transportation Research Board*, no. 2254 (2011): 170–79.

24. Juliet Jain and Glen Lyons, "The Gift of Travel Time," *Journal of Transport Geography* 16, no. 2 (2008): 81–89.

25. Louis A. Merlin, "A Portrait of Accessibility Change for Four US Metropolitan Areas," *Journal of Transport and Land Use* 10, no. 1 (2017): 309–36, http://dx.doi.org/10.5198/jtlu.2015.808.

26. David M. Levinson, Wes Marshall, and Kay Axhausen, *Elements of Access: Transport Planning for Engineers, Transport Engineering for Planners* (Sydney: Network Design Laboratory, 2017), 48.

27. Mokhtarian and Salomon, "How Derived Is the Demand for Travel?"

28. Mokhtarian and Salomon, "Mobility-Inclined Market Segments," 138.

29. Mokhtarian and Salomon, "How Derived Is the Demand for Travel?," 716–17.

30. Mokhtarian and Salomon, "Mobility-Inclined Market Segments," 138.

31. David Banister, "The Sustainable Mobility Paradigm," *Transport Policy* 15, no. 2 (2008): 73–80.

32. Banister, "Sustainable Mobility Paradigm," 75.

33. Mokhtarian, Salomon, and Redmond, "Understanding the Demand for Travel," 355; Mokhtarian and Salomon, "How Derived Is the Demand for Travel?"

34. Reid Ewing and Robert Cervero, "Travel and the Built Environment: A Meta-Analysis," *Journal of the American Planning Association* 76, no. 3 (2010): 265–94; Mark R. Stevens, "'Does Compact Development Make People Drive Less?': The Answer Is Yes," *Journal of the American Planning Association* 83, no. 1 (2016): 19–25.

35. Shima Hamidi et al., "Measuring Sprawl and Its Impacts: An Update," *Journal of Planning Education and Research* 35, no. 1 (2015): 35–50.

36. Randal O'Toole, *Gridlock: Why We're Stuck in Traffic and What to Do About It* (Washington, DC: Cato Institute, 2010), 169.

37. Mokhtarian and Salomon, "Mobility-Inclined Market Segments," 139.

38. John B. Lansing, Robert W. Marans, and Robert B. Zehner, *Planned Residential Environments* (Ann Arbor, MI: Survey Research Center, Institute for Social Research, 1970); Robert Cervero, "Jobs-Housing Balancing and Regional Mobility," *Journal of the American Planning Association* 55, no. 2 (1989): 136–50; Marlon Boarnet and Randall Crane, *Travel by Design: The Influence of Urban Form on Travel* (New York: Oxford

University Press, 2001); Transportation Research Board, *Driving and the Built Environment*; Reid Ewing et al., "Testing Newman and Kenworthy's Theory of Density and Automobile Dependence," *Journal of Planning Education and Research* 38, no. 2 (2018): 167–82.

39. Marcial H. Echenique et al., "Does Urban Form Really Matter?," *Journal of the American Planning Association* 78, no. 2 (2012): 121–37; Stevens, "Does Compact Development Make People Drive Less?," 7–18.

40. Ewing et al., "Testing Newman and Kenworthy's Theory."

41. Robert Cervero and Jin Murakami, "Effects of Built Environments on Vehicle Miles Traveled: Evidence from 370 US Urbanized Areas," *Environment and Planning A* 42, no. 2 (2010): 400–418; James McIntosh et al., "The Role of Urban Form and Transit in City Car Dependence: Analysis of 26 Global Cities from 1960 to 2000," *Transportation Research Part D* 33 (2014): 95–110.

42. CalTrans, *Guide for the Preparation of Traffic Impact Studies* (Sacramento: Department of Transportation, State of California, December 2002), 19.

43. Liam Dillon, "Want a Bike Lane in Your Neighborhood?: It's Not So Simple in California," *Los Angeles Times*, April 7, 2016, http://www.latimes.com/politics/la-pol-sac-enviro-bike-lanes-20160407-story.html.

44. According to the CEQA law, the alternative metrics could include "vehicle miles traveled, vehicle miles traveled per capita, automobile trip generation rates, or automobile trips generated."

45. Governor's Office of Planning and Research (CA), "Revised Proposal on Updates to the CEQA Guidelines on Evaluating Transportation Impacts in CEQA" (Sacramento, CA, 2016), https://www.opr.ca.gov/docs/Revised_VMT_CEQA_Guidelines_Proposal_January_20_2016.pdf.

46. Ewing and Cervero, "Travel and the Built Environment," 265–94; Dimitris Milakis, Robert Cervero, and Bert van Wee, "Stay Local or Go Regional?: Urban Form Effects on Vehicle Use at Different Spatial Scales: A Theoretical Concept and Its Application to the San Francisco Bay Area," *Journal of Transport and Land Use* 8, no. 2 (2015): 59–86.

47. Crane, "On Form versus Function," 117–26.

48. Bert van Wee, "Evaluating the Impact of Land Use on Travel Behavior: The Environment versus Accessibility," *Journal of Transport Geography* 19, no. 6 (2011): 1530.

49. Metz argues similarly that transportation-infrastructure improvement is much more likely to translate into accessibility improvement to more distant destinations than into travel-time savings. David Metz, "The Myth of Travel Time Saving," *Transport Reviews* 28, no. 3 (2008): 321–36.

2. EVOLUTION OF THE ACCESSIBILITY CONCEPT

1. Richard M. Hurd, *Principles of City Land Values* (New York: Record and Guide, 1903); Robert M. Haig, "The Assignment of Activities to Areas in Urban Regions," in *Regional Survey of New York and Its Environs* (New York: Regional Survey, 1927), 1:31–44 (originally published in *Quarterly Journal of Economics* 40 [1926]: 402–34); John Q. Stewart, "Demographic Gravitation: Evidence and Applications," *Sociometry* 11, no. 1 (1948): 31–58.

2. Stewart, "Demographic Gravitation," 57.

3. W. G. Hansen, "How Accessibility Shapes Land Use," 73–76.

4. Stewart, "Demographic Gravitation," 73.

5. Willard B. Hansen, "An Approach to the Analysis of Metropolitan Residential Extension," *Journal of Regional Science* 3, no. 1 (1961): 37–55; Donald M. Hill, "A Growth Allocation Model for the Boston Region," *Journal of the American Institute of Planners* 31, no. 2 (1965): 111–20; Nancy J. Leathers, "Residential Location and Mode of

Transportation to Work: A Model of Choice," *Transportation Research* 1, no. 2 (1967): 129–55; Knud R. Overgaard, *Traffic Estimation in Urban Transportation Planning* (Copenhagen: Danish Academy of Technical Sciences, 1966); Ross D. MacKinnon and M. John Hodgson, "Optimal Transportation Networks: A Case Study of Highway Systems," *Environment and Planning* 2, no. 3 (1970): 267–84; Alan G. Wilson, "A Family of Spatial Interaction Models, and Associated Developments," *Environment and Planning A* 3, no. 1 (1971): 1–32; M. Cordey Hayes and Alan G. Wilson, "Spatial Interaction," *Socio-Economic Planning Sciences* 5, no. 1 (1971): 73–95; David R. Ingram, "The Concept of Accessibility: A Search for an Operational Form," *Regional Studies* 5, no. 2 (1971): 101–7.

6. For example, Mei-Po Kwan, "Space-Time and Integral Measures of Individual Accessibility: A Comparative Analysis Using a Point-Based Framework," *Geographical Analysis* 30, no. 3 (1998): 191–216; Harvey J. Miller, "Measuring Space-Time Accessibility Benefits within Transportation Networks: Basic Theory and Computational Procedures," *Geographical Analysis* 31, no. 1 (1999): 1–26; Inge Thorsen and Jens P. Gitlesen, "A Simulation Approach to Studying the Sensitivity of Commuting-Flow Predictions with Respect to Specific Changes in Spatial Structure," *Environment and Planning A* 34, no. 2 (2002): 271–88.

7. Haig, "Assignment of Activities to Urban Regions," 38.

8. Haig, "Assignment of Activities to Urban Regions," 44.

9. Christopher G. Turner, "A Model Framework for Transportation and Community Plan Analysis," *Journal of the American Planning Association* 38, no. 5 (1972): 325–31.

10. Wachs and Kumagai, "Physical Accessibility as a Social Indicator," 440–41.

11. Jennifer M. Morris, Peter L. Dumble, and Marcus R. Wigan, "Accessibility Indicators for Transport Planning," *Transportation Research Part A* 13, no. 2 (1979): 91.

12. Chandra Bhat et al., *Accessibility Measures: Formulation Considerations and Current Applications*, Research Report 4938-2 (Austin: Center for Transportation Research, Bureau of Engineering Research, University of Texas, 2000), 11; David Schrank and Tim Lomax, *The 2007 Urban Mobility Report* (College Station: Texas Transportation Institute, Texas A&M University, 2007); United States Department of Transportation, *Highway Statistics 2002* (Washington, DC: Bureau of Highway Statistics, 2002); Transportation Research Board, *Highway Capacity Manual* (Washington, DC: National Academies Press, 1994); John D. Edwards Jr., *Transportation Planning Handbook* (Englewood Cliffs, NJ: Prentice-Hall, 1992).

13. Wachs and Kumagai, "Physical Accessibility as a Social Indicator," 442.

14. Jeffrey Tumlin, *Sustainable Transportation Planning: Tools for Creating Vibrant, Healthy, and Resilient Communities* (Hoboken, NJ: Wiley, 2012), 4.

15. Kevin J. Krizek, "Operationalizing Neighborhood Accessibility for Land Use–Travel Behavior Research and Regional Modeling," *Journal of Planning Education and Research* 22, no. 3 (2003): 270–87; Louis Merlin, "Measuring Community Completeness: Jobs Housing Balance, Accessibility, and Convenient Local Access to Nonwork Destinations," *Environment and Planning B* 41, no. 4 (2014): 736–56, http://doi.org/10.1068/b120010p.

16. Robert Cervero, *Paradigm Shift: From Automobility to Accessibility Planning*, Working Paper 677 (Berkeley: Institute of Urban and Regional Development, University of California, 1996), 2.

17. Susan L. Handy, *Accessibility- vs. Mobility-Enhancing Strategies for Addressing Automobile Dependence in the U.S.*, UCD-ITS-RR-02-15 (Davis: Institute of Transportation Studies, University of California, 2002).

18. Cervero, *Paradigm Shift*, 3.

19. Cervero, *Paradigm Shift*, 18.

20. Tumlin, *Sustainable Transportation Planning*, 4.

21. See the discussion of this problem in Handy, *Accessibility- vs. Mobility-Enhancing Strategies*, 3.

22. Southeast Michigan Council of Governments, *2030 Regional Transportation Plan for Southeast Michigan* (Detroit: Southeast Michigan Council of Governments, 2004), http://s3.amazonaws.com/zanran_storage/library.semcog.org/ContentPages/2455560602.pdf.

23. CalTrans, *Performance Measure for Rural Transportation Systems: Guidebook* (2006), sec. 3-1, http://www.dot.ca.gov/perf/library/pdf/RSPMGuidebook.pdf.

24. CalTrans, *Performance Measure*, sec. 4-1.

25. Boisjoly and El-Geneidy, "How to Get There?"

26. United States Code of Federal Regulations, Title 23, Highways, chap. 1, sec. 134(h)(1)(D) (2006).

27. National Research Council, *Key Transportation Indicators: Summary of a Workshop*, ed. Janet Norwood and Jamie Casey, Committee on National Statistics, Division of Behavioral and Social Sciences and Education (Washington, DC: National Academy Press, 2002), https://www.nap.edu/read/10404/chapter/4#20.

28. Randall Crane, "Counterpoint: Accessibility and Sprawl," *Journal of Transport and Land Use* 1, no. 1 (2008): 17.

3. ACCESSIBILITY IN EVERYDAY PLANNING

1. For reviews, see Geurs and van Wee, "Accessibility Evaluation"; and Handy and Niemeier, "Measuring Accessibility," 1175–94.

2. Qing Shen, "Location Characteristics of Inner-City Neighborhoods and Employment Accessibility of Low-Wage Workers," *Environment and Planning B* 25, no. 3 (1998): 345–65.

3. Joe Grengs, "Nonwork Accessibility as a Social Equity Indicator," *International Journal of Sustainable Transportation* 9, no. 1 (2015): 1–14.

4. A common form used in standard transportation planning models is the negative exponential function. Transportation Research Board, *Travel Estimation Techniques for Urban Planning*, NCHRP Report 365 (Washington, DC: National Academies Press, 1998).

5. The result will differ by the mode of travel—a car provides a far wider range of coverage than a bus for a given travel time.

6. Walter Isard, *Methods of Regional Analysis: An Introduction to Regional Science* (Cambridge, MA: MIT Press, 1960); Wilson, "Family of Spatial Interaction Models," 1–32.

7. Moshe E. Ben-Akiva and Steven E. Lerman, *Discrete Choice Analysis: Theory and Application to Predict Travel Demand*, vol. 9 (Cambridge, MA: MIT Press, 1985).

8. Joe Grengs, "Equity and the Social Distribution of Job Accessibility in Detroit," *Environment and Planning B* 39, no. 5 (2012): 785–800; Joe Grengs, "Comparing People and Places with Transportation Accessibility in Metropolitan Detroit," in *Mapping Detroit: Land, Community, and Shaping a City*, ed. June M. Thomas and Henco Bekkering (Detroit, MI: Wayne State University Press, 2015): 101–14; Grengs et al., "Intermetropolitan Comparison of Transportation Accessibility: Sorting Out Mobility and Proximity in San Francisco and Washington, D.C.," *Journal of Planning Education and Research* 29, no. 4 (2010): 427–43.

9. Grengs et al., "Intermetropolitan Comparison"; Levine et al., "Does Accessibility Require Density or Speed?: A Comparison of Fast versus Close in Getting Where You Want to Go in U.S. Metropolitan Regions," *Journal of the American Planning Association* 78, no. 2 (2012): 157–72.

10. Keith Bartholomew, Reid Ewing, and Gail Meakins, "Integrated Transportation Scenario Planning," *Environment and Planning* FHWA-HEP-10-034 (July 2010): 1–229.

11. Boisjoly and El-Geneidy "How to Get There?"; Proffitt et al., "Accessibility Planning in American Metropolitan Areas: Are We There Yet?," *Urban Studies* 15 (2017): 1–26.

12. Chicago Metropolitan Agency for Planning, *Preferred Regional Scenario* (Chicago: Chicago Metropolitan Agency for Planning, 2008); Metropolitan Planning Commission and Association of Bay Area Governments, Equity Analysis Report: Plan BayArea 2040 Final Supplemental Report (San Francisco: Metropolitan Transportation Commission and Association of Bay Area Governments, 2017), http://2040.planbayarea.org/sites/default/files/2017-07/Equity_Report_PBA%202040%20_7-2017.pdf.

13. ARC, *LCI Transportation Program* (Atlanta, GA: Atlanta Regional Commission, 2010), http://www.atlantaregional.com/land-use/livable-centers-initiative/lci-transportation-program; Envision Utah, *Wasatch Choices 2040: A Four County Land-Use and Transportation Vision* (Salt Lake City, UT: Envision Utah, 2006).

14. Kay W. Axhausen, "Accessibility Long Term Perspectives," *Journal of Transport and Land Use* 1, no. 2 (2008): 5–22; Andrew Owen, Brendan Murphy, and David M. Levinson, *Access Across America: Auto 2015*, Report CTS 16-07 (Minneapolis: Accessibility Observatory, Center for Transportation Studies, Department of Civil, Environmental, and Geo-Engineering, University of Minnesota, 2016), http://access.umn.edu/research/america/auto/2015/.

15. The method described in this chapter for land-development projects is implemented for the United States on the freely accessible website: http://aim.fau.edu.

16. Grengs et al., "Intermetropolitan Comparison of Transportation Accessibility," 427–43.

17. For details on the described procedure, see Jonathan Levine, Louis Merlin, and Joe Grengs, "Project-Level Accessibility Analysis for Land-Use Planning," *Transport Policy* 53 (2017): 107–19.

18. Arthur C. Nelson, *Planner's Estimating Guide* (Washington DC: American Planning Association Planners Press, 2004).

19. Marlon G. Boarnet and Andrew F. Haughwout, "Do Highways Matter?: Evidence and Policy Implications of Highways' Influence on Metropolitan Development" (Berkeley: University of California Transportation Center, 2000).

20. San Antonio–Bexar County Metropolitan Planning Organization, *Mobility 2040* (San Antonio, TX, 2015).

21. Louis A. Merlin, Jonathan Levine, and Joe Grengs, "Accessibility Analysis for Transportation Projects and Plans," *Transport Policy* 69 (2018): 35–48.

22. Enrica Papa et al., "Accessibility Instruments for Planning Practice: A Review of European Experiences," *Journal of Transport and Land Use* 3, no. 3 (2015): 1–20.

4. ACCESSIBILITY AND URBAN FORM

1. Reid Ewing, "Characteristics, Causes, and Effects of Sprawl: A Literature Review," *Urban Ecology* 21, no. 2 (1994): 519–35; Cervero, *Paradigm Shift*, 7.

2. Sam Staley and Adrian T. Moore, *Mobility First: A New Vision for Transportation in a Globally Competitive Twenty-First Century* (Lanham, MD: Rowman & Littlefield, 2009).

3. Owen, Murphy, and Levinson, *Access Across America: Auto 2015*.

4. See Joe Grengs, "Measuring Change in Small-Scale Transit Accessibility with Geographic Information Systems: The Cases of Buffalo and Rochester," *Transportation Research Record: Journal of the Transportation Research Board*, no. 1887 (2004): 10–17; and Owen, Murphy, and Levinson, *Access Across America: Auto 2015*.

5. See David Levinson, Bernadette Marion, and Michael Iacono, *Access to Destinations, Phase 3: Measuring Accessibility by Automobile*, Mn/DOT 2010-09 (Minneapolis: Department of Civil Engineering, University of Minnesota, 2010); Itzhak Benenson et al.,

"Public Transport versus Private Car: GIS-Based Estimation of Accessibility Applied to the Tel Aviv Metropolitan Area," *Annals of Regional Science* 47, no. 3 (2011): 499–515; Darren Scott and Mark Horner, "Examining the Role of Urban Form in Shaping People's Accessibility to Opportunities: An Exploratory Spatial Data Analysis," *Journal of Transport and Land Use* 1, no. 2 (2008): 89–119; Cynthia Chen, *The Temporal and Social Dimension of Accessibility for New York City Residents*, RF 49777-13-18 (New York: University Transportation Research Center, City College of New York, 2008); and Jianquan Q. Cheng, Luca Bertolini, and Frank le Clercq, "Measuring Sustainable Accessibility," *Transportation Research Record: Journal of the Transportation Research Board*, no. 2017 (2007): 16–25.

6. See Grengs et al., "Intermetropolitan Comparison," 427–43; Levine et al., "Does Accessibility Require Density or Speed?," 157–72; and Owen, Murphy, and Levinson, *Access Across America: Auto 2015*.

7. The complete study is available in Levine et al., "Does Accessibility Require Density or Speed?," 157–72.

8. Levine et al., "Does Accessibility Require Density or Speed?," 157–72.

9. Grengs et al., "Intermetropolitan Comparison," 427–43.

10. As explained more fully in Grengs et al., "Intermetropolitan Comparison," the transformation is accomplished by calculating a z-score for each value in a zone-to-zone travel speed matrix from metro A. This z-score matrix is then applied to the mean and standard deviation of speeds from metro B to transform the speed distribution of metro A into that of metro B.

11. Staley and Moore, *Mobility First*.

12. Claritas, Inc., *Claritas Business-Facts Methodology* (Ithaca, NY: Claritas, 2002).

13. George Galster et al., "Wrestling Sprawl to the Ground: Defining and Measuring an Elusive Concept," *Housing Policy Debate* 12, no. 4 (2001): 681–717; Bumsoo Lee, "'Edge' or 'Edgeless' Cities?: Urban Spatial Structure in US Metropolitan Areas, 1980 to 2000," *Journal of Regional Science* 47, no. 3 (2007): 479–515.

14. Trevor Thomas et al., "Not So Fast?: Examining Neighborhood-Level Effects of Traffic Congestion on Job Access, *Transportation Research Part A* 113 (2018): 529–41.

15. Reid Ewing, Rolf Pendall, and Don Chen, *Sprawl Scores for 83 Metropolitan Regions* (Washington, DC: Smart Growth America, 2002).

16. David Schrank and Tim Lomax, *2009 Urban Mobility Report* (College Station: Texas Transportation Institute, Texas A&M University, 2009).

17. Andrew Owen, Brendan Murphy, and David Levinson, *Access Across America: Transit 2015*, Report CTS 16-09 (Minneapolis: Accessibility Observatory, Center for Transportation Studies, Department of Civil, Environmental, and Geo-Engineering, University of Minnesota, 2016), http://www.cts.umn.edu/Publications/ResearchReports/reportdetail.html?id=2554.

5. THE SPECIAL CASE OF PUBLIC-TRANSPORT ACCESSIBILITY

1. Jarrett Walker, *Human Transit: How Clearer Thinking about Public Transit Can Enrich Our Communities and Our Lives* (Washington, DC: Island Press, 2012), 20.

2. Transportation Research Board, *Expanding Metropolitan Highways*.

3. Jonathan Levine et al., *Metropolitan Accessibility and Transportation Sustainability: Comparative Indicators for Policy Reform* (Final Report from EPA Agreement Number: RD-833334901-0 and FHWA Cooperative Agreement Number: DTFH61-07-H-00037, January 2011), http://hdl.handle.net/2027.42/147459.

4. Eric Jaffe, "Explaining Transit's Secret Language," *CityLab* (March 8, 2012), https://www.citylab.com/transportation/2012/03/teaching-language-transit/1438/.

5. Wait time is an important component of the total travel time in public transport, and it is related to frequency (or headways). Some studies show that wait time is on average about one-half of the expected headway when transit service is reliable, but others show that the average wait time tends to increase as a share of the expected headway, especially for headways longer than thirty minutes. Daniel B. Hess, Jeffrey Brown, and Donald Shoup, "Waiting for the Bus," *Journal of Public Transportation* 7, no. 4 (2004): 67–84.

6. Benenson et al., "Public Transport versus Private Car"; Aaron Golub, Glenn Robinson, and Brendan Nee, "Making Accessibility Analyses Accessible: A Tool to Facilitate the Public Review of the Effects of Regional Transportation Plans on Accessibility," *Journal of Transport and Land Use* 6, no. 3 (2013): 17–28; Joe Grengs, "Job Accessibility and the Modal Mismatch in Detroit," *Journal of Transport Geography* 18, no. 1 (2010): 42–54; Mark W. Horner, "Exploring Metropolitan Accessibility and Urban Structure," *Urban Geography* 25, no. 3 (2004): 264–84; Mizuki Kawabata, "Spatiotemporal Dimensions of Modal Accessibility Disparity in Boston and San Francisco," *Environment and Planning A* 41, no. 1 (2009): 183–98; Daniel B. Hess, "Access to Employment for Adults in Poverty in the Buffalo-Niagara Region," *Urban Studies* 42, no. 7 (2005): 1177–200; Thomas W. Sanchez, Qing Shen, and Zhong-Ren Peng, "Transit Mobility, Jobs Access and Low-Income Labour Participation in US Metropolitan Areas," *Urban Studies* 41, no. 7 (2004): 1313–31.

7. Grengs, "Measuring Change," 10–17; Steven E. Polzin, Ram M. Pendyala, and Sachin Navari, "Development of Time-of-Day–Based Transit Accessibility Analysis Tool," *Transportation Research Record: Journal of the Transportation Research Board*, no. 1799 (2002): 35–41.

8. One study devised a weighted average of sixteen accessibility indices calculated at each hour throughout a day between the hours of 5:00 a.m. and 9:00 p.m. Yingling Fan, Andrew E. Guthrie, and David M. Levinson, "Impact of Light Rail Implementation on Labor Market Accessibility: A Transportation Equity Perspective," *Journal of Transport and Land Use* 5, no. 3 (2012): 28–39. Others include Steven Farber, Melinda Z. Morang, and Michael J. Widener, "Temporal Variability in Transit-Based Accessibility to Supermarkets," *Applied Geography* 53 (2014): 149–59; Andrew Owen and David M. Levinson, "Modeling the Commute Mode Share of Transit Using Continuous Accessibility to Jobs," *Transportation Research Part A* 74 (2015): 110–22; and Adie Tomer et al., *Missed Opportunity: Transit and Jobs in Metropolitan America* (Washington, DC: Brookings Institution, 2011).

9. Inshu Minocha et al., "Analysis of Transit Quality of Service and Employment Accessibility for the Greater Chicago, Illinois, Region," *Transportation Research Record: Journal of the Transportation Research Board*, no. 2042 (2008): 20–29.

10. Tomer et al., *Missed Opportunity*.

11. Examples of accessibility calculations based on GTFS include Andrew Owen and David M. Levinson, *Access Across America: Transit 2014 Data* (Minneapolis: University of Minnesota Digital Conservancy, 2014), http://dx.doi.org/10.13020/D6MW2Q); and Tomer et al., *Missed Opportunity*.

12. Owen and Levinson, in "Modeling the Commute Mode Share," provide an example of allowing for an unlimited number of transfers, an approach that is possible only with fairly sophisticated technical and computer capacity. Aside from the technical challenges of accounting for transfers, because public-transport services in many regions are provided by multiple agencies that may not coordinate services, it may not be realistic to assume seamless connections across systems.

13. Transit Cooperative Research Program, *Transit Capacity and Quality of Service Manual*, 3rd ed. (Washington, DC: Transportation Research Board, 2013).

14. As defined in chapter 2, "accessibility" always refers to the ability to reach destinations; "access," by contrast, is the ability to reach the transportation system itself (in some contexts the ability to reach the system itself is also called "availability"). "Accessibility" is a commonly used but misleading term in reference to a particular transit stop or station, as a "measure of the ability of all people within a defined area to get to a specific transit station" (Transit Cooperative Research Program, *Transit Capacity and Quality of Service Manual*, 2nd ed. (Washington, DC: Transportation Research Board, 2003).

15. Arlington County, *Additional Performance Measures* (Arlington County, Virginia, 2015), 29, https://transportation.arlingtonva.us/wp-content/uploads/sites/19/2015/06/FY14_Transportation_Performance_Measures_Appendix.pdf.

16. Arlington County, *Additional Performance Measures*, 29.

17. For examples of studies using various dimensions to evaluate public-transport outcomes, see Ming Zhang, "Bus versus Rail: Meta-Analysis of Cost Characteristics, Carrying Capacities, and Land Use Impacts," *Transportation Research Record: Journal of the Transportation Research Board*, no. 2110 (2009): 87–95; Daniel B. Hess, Brian Taylor, and Allison Yoh, "Light Rail Lite or Cost-Effective Improvements to Bus Service?: Evaluating Costs of Implementing Bus Rapid Transit," *Transportation Research Record: Journal of the Transportation Research Board*, no. 1927 (2005): 22–30; Arthur C. Nelson and Joanna Ganning, *National Study of BRT Development Outcomes*, National Institute for Transportation and Communities Final Report NITC-RR-650 (Portland, OR: National Institute for Transportation and Communities, 2015), http://ppms.trec.pdx.edu/media/project_files/NITC-RR-650_ELXlH8p.pdf; Aiga Stokenberga, "Does Bus Rapid Transit Influence Urban Land Development and Property Values: A Review of the Literature," *Transport Reviews* 34, no. 3 (2014): 276–96; Sara I. Mohammad et al., "A Meta-Analysis of the Impact of Rail Projects on Land and Property Values," *Transportation Research A* 50 (2013): 158–70; Graham Currie, "Bus Transit Oriented Development—Strengths and Challenges Relative to Rail," *Journal of Public Transportation* 9, no. 4 (2006): 1–21; Milena Scherer, "Is Light Rail More Attractive to Users Than Bus Transit?: Arguments Based on Cognition and Rational Choice," *Transportation Research Record: Journal of the Transportation Research Board*, no. 2144 (2010): 11–19; Christopher Higgins, Mark Ferguson, and Pavlos Kanaroglou, "Light Rail and Land Use Change: Rail Transit's Role in Reshaping and Revitalizing Cities," *Journal of Public Transportation* 17, no. 2 (2014): 93–112; David Banister and Mark Thurstain-Goodwin, "Quantification of the Non-Transport Benefits Resulting from Rail Investment," *Journal of Transport Geography* 19, no. 2 (2011): 212–23; and Sutapa Bhattacharjee and Andrew R. Goetz, "Impact of Light Rail on Traffic Congestion in Denver," *Journal of Transport Geography* 22 (2012): 262–70. Rail transport has been evaluated for the consumer surplus it offers to passengers as a function of total passenger kilometers and fares as compared to the public investment required for its deployment. Erick Guerra, "Valuing Rail Transit: Comparing Capital and Operating Costs with Consumer Benefits," *Transportation Research Record: Journal of the Transportation Research Board*, no. 2219 (2011): 50–58.

18. Other commonly measured impacts of public transport—and dimensions on which bus and rail have been compared—are related but not equivalent to public transport's accessibility impacts. For example, public transport's impact on property values is surely a function of the accessibility benefit that public transport brings, yet property values themselves are not a full measure of the benefit that comes with accessibility improvements, since prices are affected by extraneous factors such as zoning-related supply constraints. Public-transport usage and the consumer-surplus metrics that are based on it are particularly important outcomes in that they shed direct light on public transport's performance, and any evaluation of public transport must include usage or route productivity. Though ridership volume is central to establishing the usefulness of public-transport services, it is insufficient as a metric of their quality.

19. Fan, Guthrie, and Levinson, "Impact of Light Rail Implementation," 28–39.

20. Alireza Ermagun and David Levinson, "Accessibility and Transit Performance," Minneapolis: University of Minnesota Nexus Working Papers 000129 (2015), https://conservancy.umn.edu/bitstream/handle/11299/179832/AccessibilityTransitPerformance.pdf?sequence=1&isAllowed=y.

21. We gratefully acknowledge the innovative dataset compiled at the University of Minnesota. Ermagun and Levinson, "Accessibility and Transit Performance" (2015).

22. Levine et al., "Does Accessibility Require Density or Speed?," 157–72.

23. Robert Cervero, *The Transit Metropolis: A Global Inquiry* (Washington, DC: Island Press, 1998).

24. National Transit Database, *NTD Annual Database*, https://www.transit.dot.gov/ntd/ntd-data. Fixed guideway is a category whose service, at 91 percent of VRM, is dominantly but not exclusively in rail. Other service types in this NTD category include some BRT, fixed catenary, ferry, or any mode that uses a separate right-of-way for the exclusive use of public transport. Non-fixed guideway refers to public transport in mixed-traffic rights-of-way, and includes the modes of bus, demand response, commuter bus, and the remainder of BRT. A large majority (91 percent) of service in the non-fixed-guideway category is provided by buses.

25. National Transit Database, *NTD Annual Database*. Fixed-guideway service—the "rail" category in this analysis includes 6.5 million VRM in BRT, or 1 percent of all fixed-guideway service. Non-fixed-guideway service includes 0.6 million VRM in BRT, accounting for 0.03 percent of non-fixed-guideway VRM. While high-standard BRT is based on dedicated lanes, many self-identified BRT systems in the United States include segments (or in some cases entire lines) of bus service in mixed traffic. Institute for Transportation and Development Policy, *The Bus Rapid Transit Standard* (New York, 2016), https://www.itdp.org/library/standards-and-guides/the-bus-rapid-transit-standard/the-scorecard/.

26. Accessibility Observatory, Center for Transportation Studies, University of Minnesota, Access Across America: Transit 2014 Data, 2014, http://access.umn.edu/data/datasets/; National Transit Database; Minnesota Population Center, *National Historical Geographic Information System*, Version 2.0 (Minneapolis: University of Minnesota, 2011), http://www.nhgis.org; U.S. Census Bureau, *LODES Data* (Longitudinal-Employer Household Dynamics Program, 2015), http://lehd.ces.census.gov/data/lodes/.

27. This is a weighted accessibility score for each metropolitan area based on the average number of jobs that are reachable by transit within 10, 20, 30, 40, 50, and 60 minutes; within the weighting scheme, jobs that take longer to reach by public transport count for less than jobs that are reachable in less time (for a more detailed description of the methods for the accessibility score, see Owen and Levinson, *Access Across America: Transit 2014 Data*. This creates a metric similar to gravity-based accessibility, with the weighting of the various isochrones playing the same role as the impedance parameter. Levine et al., "Does Accessibility Require Density or Speed?," 157–72; Grengs et al., "Intermetropolitan Comparison," 427–43.

28. A metropolitan area will have a higher fit if it has higher-density residential development in areas with superior public-transport accessibility. The Fit Index varies from highs of around 0.35 for the Milwaukee, Minneapolis, and New York metropolitan regions to lows of around 0.02 for the Dallas, Orlando, and Raleigh metropolitan regions. A higher Gini Index indicates that the distribution of public-transport accessibility is concentrated among a small percentage of the population, and therefore that public-transport accessibility is spread less evenly. A public-transport system in which all resources are concentrated on just one line serving a highly populated corridor would lead to a highly uneven distribution of public-transport accessibility and a high Gini Index; a system that spreads public-transport service more evenly across many lines reaching all

corners of the metropolitan region would lead to a lower Gini Index. The Gini Index is high for all regions of this study, indicating an uneven intrametropolitan distribution of public-transport accessibility. The index ranges from lows around 0.60 for metropolitan areas including San Jose, Salt Lake City, and Riverside to highs above 0.84 for the metropolitan areas with the most uneven distribution of public-transport accessibility: Pittsburgh, Nashville, and Atlanta.

29. Several regression models (not shown here) were constructed to test the contribution of rail and bus modes to job accessibility. First, a model was estimated for all forty-six metropolitan areas. New York exerted high leverage over the regression, which raised the concern that this regression was unduly influenced by this extreme case. To ensure that results were not an artifact of this leverage, models presented here were constructed without New York.

30. Jonathan Levine, "Is Bus versus Rail Investment a Zero-Sum Game?: The Misuse of the Opportunity-Cost Concept," *Journal of the American Planning Association* 79, no. 1 (2013): 5–15.

31. Because of data limitations, the shift from average to 85th-percentile accessibility necessitated moving from job accessibility weighted across all time bands to the thirty-minute time band alone. To test the impact of this shift, models were estimated using the mean thirty-minute time band and compared with the average weighted accessibility. The similarity between these models suggests that differences between models A and B on the one hand and C and D on the other are principally attributable to the shift from the mean resident to the 85th-percentile resident, rather than the shift from average weighted accessibility to thirty-minute accessibility.

32. Charting high-Gini and low-Gini metropolitan areas of similar size and with similar levels of total public-transport service (VRM per capita) shows that the accessibility difference between two metros at the 85th percentile is considerably greater than at the mean of the accessibility distribution. For example, Salt Lake City (Gini=0.61) has a broader spread of public transport than Charlotte (Gini=0.84), and public-transport accessibility is much higher in Salt Lake City as compared to Charlotte at the 85th percentile.

33. Levine, "Is Bus versus Rail Investment a Zero-Sum Game?," 5–15.

34. United States General Accounting Office, *Bus Rapid Transit Shows Promise* (Washington DC: U.S. General Accounting Office, 2001). Not only are rail capital costs much higher than buses, so too are busways. The GAO reports busway capital costs to be over nineteen times the capital costs of buses on arterials.

35. Zhang, "Bus Versus Rail," 87–95.

36. Calculations based on National Transit Database for 2013 data.

37. Robert Cervero, "Bus Rapid Transit (BRT): An Efficient and Competitive Mode of Public Transport," Working Paper 2013-01 (Berkeley, CA: Institute of Urban and Regional Development, 2013), http://iurd.berkeley.edu/wp/2013-01.pdf.

38. National Cooperative Highway Research Program, *Default Values for Highway Capacity and Level of Service Analyses*, NCHRP Report 599 (Washington DC: National Cooperative Highway Research Program, 2008).

39. Levine, "Is Bus Versus Rail Investment a Zero-Sum Game?," 5–15.

40. Levine, *Zoned Out*.

6. ACCESSIBILITY IN SOCIAL-EQUITY EVALUATION

1. Devajyoti Deka, "Social and Environmental Justice Issues in Urban Transportation," in *The Geography of Urban Transportation*, ed. Susan Hanson and Genevieve Giuliano, 3rd ed. (New York: Guilford Press, 2004), 332–55; David J. Forkenbrock and Lisa A. Schweitzer, "Environmental Justice in Transportation Planning," *Journal of the American Planning Association* 65, no. 1 (1999): 96–111; Alan Hay, "Equity and Welfare

in the Geography of Public Transport Provision," *Journal of Transport Geography* 1, no. 2 (1993): 95–101; David C. Hodge, "My Fair Share: Equity Issues in Urban Transportation," in *The Geography of Urban Transportation*, ed. Susan Hanson, 2nd ed. (New York: Guilford Press, 1995), 359–75; Lisa Schweitzer and Abel Valenzuela Jr., "Environmental Injustice and Transportation: The Claims and the Evidence," *Journal of Planning Literature* 18, no. 4 (2004): 383–98; Brian D.Taylor and Alexandra Tassiello Norton, "Paying for Transportation: What's a Fair Price?," *Journal of Planning Literature* 24, no. 1 (2009): 22–36.

2. Transportation Research Board, *Expanding Metropolitan Highways*.

3. Robert D Bullard, Angel O. Torres, and Glenn S. Johnson, eds., *Sprawl City: Race, Politics, and Planning in Atlanta* (Washington, DC: Island Press, 2000); Rolf Pendall, "Do Land Use Controls Cause Sprawl?," *Environment and Planning B* 26 (1999): 555–71; Gregory D. Squires and Charis E. Kubrin, "Privileged Places: Race, Uneven Development and the Geography of Opportunity in Urban America," *Urban Studies* 42, no. 1 (2005): 47–68.

4. Jago Dodson et al., "Travel Behavior Patterns of Different Socially Disadvantaged Groups: Analysis of Household Travel Survey Data for a Dispersed Metropolitan Area," *Transportation Research Record: Journal of the Transportation Research Board*, no. 2163 (2010): 24–31; David J. Forkenbrock and Glen E. Weisbrod, *Guidebook for Estimating the Social and Economic Effects of Transportation Projects*, NCHRP Report 456 (Washington, DC: National Academy Press, 2001); Genevieve Giuliano, "Travel, Location and Race/Ethnicity," *Transportation Research Part A* 37, no. 4 (2003): 351–72; Ibipo Johnston-Anumonwo, "Racial Differences in the Commuting Behavior of Women in Buffalo, 1980–1990," *Urban Geography* 16, no. 1 (1995): 23–45.

5. John Pucher and John L. Renne, "Socioeconomics of Urban Travel: Evidence from the 2001 NHTS," *Transportation Quarterly* 57, no. 3 (2003): 49–77; Brian D. Taylor and Paul M. Ong, "Spatial Mismatch or Automobile Mismatch?: An Examination of Race, Residence, and Commuting in U.S. Metropolitan Areas," *Urban Studies* 32, no. 9 (1995): 1537–57.

6. Elizabeth Deakin, "Equity and Environmental Justice in Sustainable Transportation: Toward a Research Agenda," in *Institutions and Sustainable Transport: Regulatory Reform in Advanced Economies*, ed. Piet Rietveld and Roger R. Stough (Cheltenham, UK: Edward Elgar, 2007), 51–69; Alex Karner and Deb A. Niemeier, "Civil Rights Guidance and Equity Analysis Methods for Regional Transportation Plans: A Critical Review of Literature and Practice," *Journal of Transport Geography* 33 (2013): 126–34; Scott G. Mills and Sieglinde K. Neuhauser, "Quantitative Methods for Environmental Justice Assessment of Transportation," *Risk Analysis* 20, no. 3 (2000): 377–84; Schweitzer and Valenzuela, "Environmental Injustice and Transportation," 383–98.

7. Social Exclusion Unit, *Making the Connections: Final Report on Transport and Social Exclusion* (London: Office of the Deputy Prime Minister, 2003); John Farrington and Conor Farrington, "Rural Accessibility, Social Inclusion and Social Justice: Towards Conceptualisation," *Journal of Transport Geography* 13, no. 1 (2005): 1–12; John Preston and Fiona Rajé, "Accessibility, Mobility and Transport-Related Social Exclusion," *Journal of Transport Geography* 15, no. 3 (2007): 151–60; Bert van Wee and Karst T. Geurs, "Discussing Equity and Social Exclusion in Accessibility Evaluations," *European Journal of Transport and Infrastructure Research* 11, no. 4 (2011): 350–67.

8. Social Exclusion Unit, *Making the Connections*.

9. Karen Lucas, "Transport and Social Exclusion: Where Are We Now?," *Transport Policy* 20 (2012): 105–13.

10. Lucas, "Transport and Social Exclusion," 112.

11. Shannon Cairns, Jessica Greig, and Martin Wachs, *Environmental Justice and Transportation: A Citizen's Handbook*, UCTC No. 620 (Berkeley: University of California

Transportation Center, 2003); Thomas W. Sanchez and Marc Brenman, *The Right to Transportation: Moving to Equity* (Chicago: American Planning Association, 2007).

12. Karner and Niemeier, "Civil Rights Guidance," 126–34; Thomas W. Sanchez, Rich Stolz, and Jacinta S. Ma, *Moving to Equity: Addressing Inequitable Effects of Transportation Policies on Minorities* (Cambridge, MA: Civil Rights Project at Harvard University, 2003).

13. Cambridge Systematics, *Technical Methods to Support Analysis of Environmental Justice Issues* (Cambridge, MA: National Cooperative Highway Research Program, 2002), 1.

14. Halden, "Use and Abuse of Accessibility Measures"; Karen Lucas, "A Critical Assessment of Accessibility Planning for Social Inclusion," in *Accessibility Analysis and Transport Planning: Challenges for Europe and North America*, ed. Karst T. Geurs, Kevin J. Krizek, and Aura Reggiani (Cheltenham, UK: Edward Elgar, 2012), 245–62; Lucas, "Transport and Social Exclusion," 105–13.

15. Federal Transit Administration, *FTA Circular 4702.1A: Title VI and Title VI–Dependent Guidelines for Federal Transit Administration Recipients* (Washington, DC: U.S. Department of Transportation, 2007); Federal Transit Administration, *FTA Circular 4702.1B: Title VI Requirements and Guidelines for Federal Transit Administration Recipients* (Washington, DC: U.S. Department of Transportation, 2012).

16. Federal Transit Administration, *FTA Circular 4702.1A*, VI-1.

17. Cambridge Systematics, *Technical Methods to Support Analysis*; Karner and Niemeier, "Civil Rights Guidance," 126–34.

18. Michael W. Steinberg, "Making Sense of Environmental Justice," *Forum for Applied Research and Public Policy* 15, no. 3 (2000): 82–89.

19. Mid-America Regional Council, *Environmental Justice Analysis* (Kansas City, MO: Mid-America Regional Council, 2013).

20. Forkenbrock and Weisbrod, *Guidebook for Estimating the Social and Economic Effects.*

21. Cambridge Systematics, *Technical Methods to Support Analysis*; Kevin Manaugh, Madhav G. Badami, and Ahmed M. El-Geneidy, "Integrating Social Equity into Urban Transportation Planning: A Critical Evaluation of Equity Objectives and Measures in Transportation Plans in North America," *Transport Policy* 37 (2015): 167–76; Charles Purvis, "Data and Analysis Methods for Metropolitan-Level Environmental Justice Assessment," *Transportation Research Record: Journal of the Transportation Research Board*, no. 1756 (2001): 15–21.

22. Boston Region Metropolitan Planning Organization, "Transportation Equity," in *Long-Range Transportation Plan of the Boston Region Metropolitan Planning Organization* (Boston, MA: Boston Region Metropolitan Planning Organization, 2015), 7-1–7-17; Metropolitan Transportation Commission, *Transportation 2035 Plan for the San Francisco Bay Area: Equity Analysis Report* (Oakland, CA: Metropolitan Transportation Commission, 2009); Mid-Ohio Regional Planning Commission, "Appendix D: Environmental Justice Technical Analysis," in *2012–2035 Metropolitan Transportation Plan* (Columbus: Mid-Ohio Regional Planning Commission, 2012).

23. Mid-Ohio Regional Planning Commission, *2012–2035 Metropolitan Transportation Plan*, 37.

24. Jonathan Levine, "Urban Transportation and Social Equity: Transportation-Planning Paradigms That Impede Policy Reform," in *Policy, Planning, and People: Promoting Justice in Urban Development*, ed. Naomi Carmon and Susan S. Fainstein (Philadelphia: University of Pennsylvania Press, 2013), 141–60.

25. Wachs and Kumagai, "Physical Accessibility," 437–56.

26. Karel Martens, "Justice in Transport as Justice in Accessibility: Applying Walzer's 'Spheres of Justice' to the Transport Sector," *Transportation* 39, no. 6 (2012): 1035–53; Martens, *Transport Justice*; Preston and Rajé, "Accessibility, Mobility," 151–60.

27. John Black and Chatherine M. Conroy, "Accessibility Measures and the Social Evaluation of Urban Structure," *Environment and Planning A* 9 (1977): 1013–31; Wachs and Kumagai, "Physical Accessibility," 437–56.

28. Todd Litman, "Evaluating Transportation Equity," *World Transport Policy and Practice* 8, no. 2 (2002): 50–65; Taylor and Tassiello Norton, "Paying for Transportation," 22–36.

29. Nicole Foth, Kevin Manaugh, and Ahmed M. El-Geneidy, "Towards Equitable Transit: Examining Transit Accessibility and Social Need in Toronto, Canada, 1996–2006," *Journal of Transport Geography* 29 (2013): 1–10; Martens, "Justice in Transport," 1035–53; Karel Martens, Aaron Golub, and Glenn Robinson, "A Justice-Theoretic Approach to the Distribution of Transportation Benefits: Implications for Transportation Planning Practice in the United States," *Transportation Research Part A* 46, no. 4 (2012): 684–95; Alan T. Murray and Rex Davis, "Equity in Regional Service Provision," *Journal of Regional Science* 41, no. 4 (2001): 577–600.

30. Cambridge Systematics, *Technical Methods to Support Analysis.*

31. Deakin, "Equity and Environmental Justice"; Karner and Niemeier, "Civil Rights Guidance," 126–34.

32. Martens, "Justice in Transport," 1048.

33. Benenson et al., "Public Transport versus Private Car"; Erin M. Ferguson et al., "Incorporating Equity into the Transit Frequency-Setting Problem," *Transportation Research Part A* 46 no. 1 (2012): 190–99; Kawabata, "Spatiotemporal Dimensions," 183–98; Rebecca C. W. Kwok and Anthony G. O. Yeh, "The Use of Modal Accessibility Gap as an Indicator for Sustainable Transport Development," *Environment and Planning A* 36 (2004): 921–36. One of the earliest recommendations for minimizing this difference was based on the view that sustainability planning requires prioritizing public transport accessibility over private transport. Kwok and Yeh, "Use of Modal Accessibility Gap," 925. Focusing on transit's accessibility relative to the car has been promoted as a useful approach to addressing the social and environmental costs of high levels of automobile dependence and rectifying auto-oriented urban spatial structure. Kawabata, "Spatiotemporal Dimensions," 183–98; Mizuki Kawabata and Qing Shen, "Job Accessibility as an Indicator of Auto-Oriented Urban Structure: A Comparison of Boston and Los Angeles with Tokyo," *Environment and Planning B* 33, no. 1 (2006): 115–30; Mizuki Kawabata and Qing Shen, "Commuting Inequality between Cars and Public Transit: The Case of the San Francisco Bay Area, 1990–2000," *Urban Studies* 44, no. 9 (2007): 1759–80.

34. Ferguson et al., "Incorporating Equity," 192.

35. Martens, Golub, and Robinson, "Justice-Theoretic Approach," 689.

36. Grengs, "Job Accessibility," 42–54; Kate Lowe and Joe Grengs, "Private Donations for Public Transit: The Equity Implications of Detroit's Public-Private Streetcar," *Journal of Planning Education and Research* (March 13, 2018), https://doi.org/10.1177/0739456X18761237; Jen Nelles, "Regionalism Redux: Exploring the Impact of Federal Grants on Mass Public Transit Governance and Political Capacity in Metropolitan Detroit," *Urban Affairs Review* 49, no. 2 (March 1, 2013): 220–53.

37. Evelyn Blumenberg and Michael Manville, "Beyond the Spatial Mismatch: Welfare Recipients and Transportation Policy," *Journal of Planning Literature* 19, no. 2 (2004): 186; see also Grengs, "Job Accessibility," 42–54.

38. Shared-Use Mobility Center, *Shared-Use Mobility Reference Guide* (Chicago: Shared-Use Mobility Center, 2015), http://sharedusemobilitycenter.org/wp-content/uploads/2015/09/SharedUseMobility_ReferenceGuide_09.25.2015.pdf.

39. Mark Garrett and Brian Taylor, "Reconsidering Social Equity in Public Transit," *Berkeley Planning Journal* 13 (1999): 6–27; Martin Wachs and Brian D. Taylor, "Can Transportation Strategies Help Meet the Welfare Challenge?," *Journal of the American Planning Association* 64, no. 1 (1998): 15–19.

40. Deakin, "Equity and Environmental Justice"; Aaron Golub, Richard A. Marcantonio, and Thomas W. Sanchez, "Race, Space, and Struggles for Mobility," *Urban Geography* 34, no. 5 (2013): 699–728; Joe Grengs, "Community-Based Planning as a Source of Political Change: The Transit Equity Movement of Los Angeles' Bus Riders Union," *Journal of the American Planning Association* 68, no. 2 (2002): 165–78; Karner and Niemeier, "Civil Rights Guidance," 126–34; Eric Mann, "Confronting Transit Racism in Los Angeles," in *Just Transportation: Dismantling Race and Class Barriers to Mobility*, ed. Robert D. Bullard and Glenn S. Johnson (Gabriola Island, BC: New Society, 1997), 68–83; Sanchez and Brenman, *Right to Transportation*; Transit Cooperative Research Program, *Civil Rights Implications of the Allocation of Funds between Bus and Rail*, Legal Research Digest 27 (Washington, DC: Transportation Research Board, 2008).

41. Edward L. Glaeser, Matthew E. Kahn, and Jordan Rappaport, "Why Do the Poor Live in Cities: The Role of Public Transportation," *Journal of Urban Economics* 63, no. 1 (2008): 1–24; Grengs, "Job Accessibility," 42–54.

42. The quartiles of transit accessibility are defined internally to each metropolitan area. The top quartile of Kansas City, for example, would represent much lower accessibility than the top quartile of New York City. This helps indicate the capacity or desire of transit-dependent households to select the transit-accessible areas within their own metropolitan region.

43. This is consistent with the pioneering work by the Center for Neighborhood Technology and the Center for Transit-Oriented Development in creating what they refer to as the "H + T Affordability Index," which in part measures the savings that derive from living in transit-rich neighborhoods.

44. William J. Clinton, *Executive Order No. 12898: Federal Actions to Address Environmental Justice in Minority Populations and Low-Income Populations, Federal Register* 59:7629 (1994); Federal Transit Administration, *FTA Circular 4702.1A*.

45. Pucher and Renne, "Socioeconomics of Urban Travel," 49–77.

46. Although the data and steps required are not outlined here, they can be found elsewhere: Grengs, "Equity and the Social Distribution," 785–800; Grengs, "Nonwork Accessibility," 1–14; Levine et al., *Metropolitan Accessibility*.

47. Grengs, "Job Accessibility," 42–54; Lingqian Hu, "Job Accessibility of the Poor in Los Angeles," *Journal of the American Planning Association* 81, no. 1 (2015): 30–45.

48. Laurent Gobillon, Harris Selod, and Yves Zenou, "The Mechanisms of Spatial Mismatch," *Urban Studies* 44, no. 12 (2007): 2401–27; Elvin K. Wyly and Daniel J. Hammel, "Gentrification, Segregation, and Discrimination in the American Urban System," *Environment and Planning A* 36, no. 7 (2004): 1215–41.

49. Several cases are notable for unusual patterns (although the figures are not shown for brevity). First, some metropolitan regions have an unusually high disparity between low-income households and high-income households, including Baltimore, Cincinnati, Los Angeles, Memphis, and Philadelphia. Second, several regions show a distinctly low disparity between low- and high-income households, including Atlanta, Chicago, Minneapolis–St. Paul, and Seattle.

50. Pucher and Renne, "Socioeconomics of Urban Travel," 49–77.

51. Harry Holzer, "The Spatial Mismatch Hypothesis: What Has the Evidence Shown?," *Urban Studies* 28, no. 1 (1991): 105–22; John F. Kain, "The Spatial Mismatch Hypothesis: Three Decades Later," *Housing Policy Debate* 3, no. 2 (1992): 371–460; Douglas S. Massey and Nancy A. Denton, *American Apartheid: Segregation and the Making of the Underclass* (Cambridge, MA: Harvard University Press, 1993); Valerie Preston and Sara McLafferty, "Spatial Mismatch Research in the 1990s: Progress and Potential," *Papers in Regional Science* 78 (1999): 387–402.

52. Xavier de Souza Briggs, ed., *The Geography of Opportunity: Race and Housing Choice in Metropolitan America* (Washington, DC: Brookings Institution, 2005); Peter Dreier, John H. Mollenkopf, and Todd Swanstrom, *Place Matters: Metropolitics for the Twenty-First Century*, 2nd ed. (Lawrence,: University Press of Kansas, 2004).

53. Accessibility models that account for job competition would likely diminish the benefits of central location to employment. Qing Shen, "Location Characteristics," 345–65.

54. William Alonso, "A Theory of the Urban Land Market," in *Readings in Urban Economics*, ed. Matthew Edel and Jerome Rothenberg (New York: Macmillan, 1972): 104–11.

7. NONWORK ACCESSIBILITY

1. United States Department of Transportation, *2009 National Household Travel Survey* (Washington, DC: Federal Highway Administration, 2011), 25.

2. United States Department of Transportation, *2009 National Household Travel Survey*.

3. United States Department of Labor, *Economic News Release: Employee Tenure Summary* (Washington DC: Bureau of Labor Statistics, 2010), http://www.bls.gov.proxy.lib.umich.edu/news.release/tenure.nr0.htm; Jason Schachter, *Geographical Mobility: March 1999 to March 2000*, Current Population Reports P20–538 (Washington, DC: U.S. Census Bureau, 2001).

4. Michael J. Widener et al., "Using Urban Commuting Data to Calculate a Spatiotemporal Accessibility Measure for Food Environment Studies," *Health & Place*, no. 21 (2013): 1–9; Wei Luo and Fahui Yang, "Measures of Spatial Accessibility to Healthcare in a GIS Environment: Synthesis and a Case Study in the Chicago Region," *Environment and Planning B*, no. 3 (2003): 865–84.

5. Peter Gordon, Ajay Kumar, and Harry W. Richardson, "Beyond the Journey to Work," *Transportation Research Part A* 22, no. 6 (1988): 419.

6. Kang-Rae Ma and David Banister, "Excess Commuting: A Critical Review," *Transport Reviews* 26, no. 6 (2006): 763–64.

7. Muhammad A. Habib and Eric J. Miller, "Reference-Dependent Residential Location Choice Model within a Relocation Context," *Transportation Research Record: Journal of the Transportation Research Board*, no. 2133 (2009): 92–99; Amaya Vega and Aisling Reynolds-Feighan, "A Methodological Framework for the Study of Residential Location and Travel-to-Work Mode Choice under Central and Suburban Employment Destination Patterns," *Transportation Research Part A* 43, no. 4 (2009): 401–19; Joseph Prashker, Yoram Shiftan, and Pazit Hershkovitch-Sarusi, "Residential Choice Location, Gender and the Commute Trip to Work in Tel Aviv," *Journal of Transport Geography* 16, no. 5 (2008): 332–41; Diana Mok, "Do Two-Earner Households Base Their Choice of Residential Location on Both Incomes?," *Urban Studies* 44, no. 4 (2007): 723–50; Patricio E. Pérez, Francisco J. Martínez, and Juan de Dios Ortúzar, "Microeconomic Formulation and Estimation of a Residential Location Choice Model: Implications for the Value of Time," *Journal of Regional Science* 43, no. 4 (2003): 771–89; Srour, Kockelman, and Dunn, "Accessibility Indices."

8. Lee et al., "Reexamining the Influence of Work and Nonwork Accessibility," 913–30.

9. Jason Chen, Cynthia Chen, and Harry Timmermans, "Accessibility Trade-Offs in Household Residential Location Decisions," *Transportation Research Record: Journal of the Transportation Research Board*, no. 2077 (2008): 71–79.

10. Lee et al., "Reexamining the Influence of Work and Nonwork Accessibility," 913–30; Chen et al., "Accessibility Trade-Offs," 71–79.

11. Dick Ettema and Harry Timmermans, "Space-Time Accessibility under Conditions of Uncertain Travel Times: Theory and Numerical Simulations," *Geographical Analysis* 39, no. 2 (2007): 217–40.

12. Philippe Apparicio and Anne-Marie Séguin, "Measuring the Accessibility of Services and Facilities for Residents of Public Housing in Montreal," *Urban Studies* 43, no. 1 (2006): 187–211.

13. Juan Carlos Martin and Aura Reggiani, "Recent Methodological Developments to Measure Spatial Interaction: Synthetic Accessibility Indices Applied to High-Speed Train Investments," *Transport Reviews* 27, no. 5 (2007): 551–71.

14. Torsten Hägerstrand, "What About People in Regional Science?," *Papers in Regional Science* 24, no. 1 (1970): 6–21; Mei-Po Kwan and Joe Weber, "Scale and Accessibility: Implications for the Analysis of Land Use–Travel Interaction," *Applied Geography* 28, no. 2 (2008): 110–23.

15. United States Department of Transportation, *2001 National Household Travel Survey* (Washington, DC: Federal Highway Administration and Bureau of Transportation Statistics, 2003).

16. The method used to calculate the coefficient of impedance is detailed in Levine et al., "Does Accessibility Require Density or Speed?," 157–72, 161. For work travel, the best-fitting regression is: estimated $\beta = 0.109^{*}\exp(-3.52^{*}10^{-8}.^{*}\text{Population})$. For nonwork travel, the best-fitting regression is: estimated $\beta = 0.24^{*}\exp(-3.52^{*}10^{-8}.^{*}\text{Population})$.

17. See Jianxin Gong, "Clarifying the Standard Deviational Ellipse," *Geographical Analysis* 34, no. 2 (2002): 155–67. Median ellipse sizes across thirty-eight metropolitan regions studied: work locations, 1,482 km^2; nonwork locations, 1642 km^2; population, 1845 km^2.

18. Levine, *Zoned Out*; Daniel G. Chatman, "Residential Choice, the Built Environment, and Nonwork Travel: Evidence Using New Data and Methods," *Environment and Planning A* 41, no. 5 (2009): 1072–89.

19. Grengs, "Nonwork Accessibility," 1–14.

CONCLUSION

1. John Stuart Mill, *On Liberty* (1859; repr., New York: Penguin Press, 1985).

2. Andrew Monschein and Brian D. Taylor, "Is Traffic Congestion Overrated?: Examining the Highly Variable Effects of Congestion on Travel and Accessibility," *Journal of Transport Geography* 64 (2017): 65–76.

3. Jeffrey Gutman and Adie Tomer, *Developing a Common Narrative on Urban Accessibility: Overview*, Moving to Access (Washington, DC: Brookings, 2016), https://www.brookings.edu/wp-content/uploads/2017/01/overview-digital.pdf.

4. Chang-Hee Christine Bae, "Air Quality and Travel Behavior: Untying the Knot," *Journal of the American Planning Association* 59, no. 1 (1993): 65–74; Martin Wachs, "Learning from Los Angeles: Transport, Urban Form, and Air Quality," *Transportation* 20, no. 4 (1993): 329–54.

5. J. Meyer et al., "Autonomous Vehicles: The Next Jump in Accessibilities?," *Research in Transportation Economics* 62 (2017): 80–90.

6. For example, Adam Lovelady, "MPOs and the Integration of Transportation and Land Use Planning," *Virginia Environmental Law Journal* 27, no. 3 (2009): 275–320; Marianne Hatzopoulou and Eric J. Miller, "Institutional Integration for Sustainable Transportation Policy in Canada," *Transport Policy* 15, no. 3 (2008): 149–62; Marco Te Brömmelstroet and Luca Bertolini, "Integrating Land Use and Transport Knowledge in Strategy-Making," *Transportation* 37, no. 1 (2010): 85–104.

7. Schrank et al., *2015 Urban Mobility Scorecard.*

8. Transportation Research Board, *Highway Capacity Manual: A Guide for Multimodal Mobility Analysis*, 6th ed. (Washington, DC: Transportation Research Board, 2016); American Association of State Highway and Transportation Officials and Knovel, *A Policy on Geometric Design of Highways and Streets*, 6th. ed. (Washington, DC: American Association of State Highway and Transportation Officials, 2011); Institute of Transportation Engineers, *Traffic Engineering Handbook* (Hoboken, NJ: Wiley, 2016).

9. Rolf Pendall, "Local Land Use Regulation and the Chain of Exclusion," *Journal of the American Planning Association* 66, no. 2 (2000): 125–42.

10. Judith E. Innes, "Information in Communicative Planning," *Journal of the American Planning Association* 68 (1998): 52–63.

11. Institute of Transportation Engineers, *Transportation Impact Analyses*, 99.

12. George Edward Moore, *Principia Ethica*, ed. Thomas Baldwin (Cambridge: Cambridge University Press, 1993); Bent Flyvbjerg, *Making Social Science Matter*.

13. "Transportation Concurrency," chap. 12.24 of the City of Snoqualmie (Washington) Municipal Code, http://www.codepublishing.com/WA/Snoqualmie/.

14. "Guidelines for Traffic Impact Analysis (Zoning Petitions)," Wake County, North Carolina, accessed February 1, 2019, http://www.wakegov.com/planning/zoning/Documents/guidelines_for_traffic_impact_analysisUDO3.pdf.

15. Bannock Planning Organization, "Traffic Impact Study Guidelines" (produced for the City of Pocatello, City of Chubbuck, and Bannock County (Idaho), 2006), http://bannockplanning.org/wp-content/uploads/trafficimpactstudy.pdf.

16. Bannock Planning Organization, "Traffic Impact Study Guidelines."

17. Mid-Region Council of Governments and Mid-Region Metropolitan Planning Organization, "A Profile in Congestion (2012): The 30 Most Congested Corridors in the Albuquerque Metropolitan Planning Area," https://www.mrcog-nm.gov/images/stories/pdf/transportation/CMP/2012-cmp-profile.pdf.

18. Wilson and Company, "Interchange Access Change Request" (Wilson and Company, unpublished memo, 2013), 28. The preferred project exhibited higher LOS ratings than the no-build alternative, ranging between C and F, for freeway, weave junction, and intersection analysis. Two out of twenty-four ramp junctions measured received a score of B, with the remaining twenty-two receiving scores of F, D, or C.

19. Wilson and Company, "Interchange Access Change Request," 36–45.

20. Mid-Region Council of Governments, *Economic Impact Report* (Albuquerque, NM: 2011), 3, 5.

21. *Karsch v. Camden County Board of Adjustments*, 302 S.W.3d 754 (M.O. App. 2010), 11.

22. *Karsch v. Camden County*, pdf p. 11.

23. "Mayor Newsom Reaffirms Commitment to City's Bike Plan," *US States News*, November 8, 2006.

24. San Francisco Planning Department, "San Francisco Bicycle Plan Environmental Impact Report," vol. 1, V.A. 1-1, 2009.

25. San Francisco Planning Department, "San Francisco Bicycle Plan," V.A. 2-6, 240.

26. Boisjoly and El-Geneidy, "How to Get There?"; David G. Proffitt et al., "Accessibility Planning," 1–26.

27. Geneviève Boisjoly and Ahmed El-Geneidy, *Measuring Performance: Accessibility Metrics in Metropolitan Regions around the World* (Washington, DC: Brookings, 2017); Boisjoly and El-Geneidy, "How to Get There?," 38–50; Proffitt, et al., "Accessibility Planning," 1–26.

28. Geneviève Boisjoly and Ahmed M. El-Geneidy, "The Insider: A Planners' Perspective on Accessibility," *Journal of Transport Geography* 64 (2017): 33–43.

29. Boisjoly and El-Geneidy, "Measuring Performance."

30. Puget Sound Regional Council, *Transportation 2040 Plan* (Seattle, WA: Puget Sound Regional Council, 2010).

31. Puget Sound Regional Council, *Transportation 2040 Plan*, G-1.

32. Puget Sound Regional Council, *Transportation 2040 Plan*, 65, 81.

33. Puget Sound Regional Council, *Transportation 2040 Plan*, 15.

34. Chicago Metropolitan Agency for Planning, *Preferred Regional Scenario*; Chicago Metropolitan Agency for Planning, *GO TO 2040: Comprehensive Regional Plan* (Chicago: Chicago Metropolitan Agency for Planning, 2010).

35. Chicago Metropolitan Agency for Planning, *GO TO 2040*, 15.

36. Eric Sundquist, "Virginia DOT's SMART Scale Project Prioritization Process" (paper presented at the *Transportation Research Board 96th Annual Meeting*, Washington DC, January 8–12, 2017).

37. Virginia Department of Transportation, *Smart Scale Technical Guide* (Richmond: Virginia Department of Transportation, 2016).

38. Tamara Rollison and Brian Coy, "Gov. McAuliffe Announces Virginia's First Prioritization Process Scores Nearly 300 Proposed Transportation Projects," Virginia Department of Transportation, January 19, 2016, http://www.virginiadot.org/newsroom/statewide/2016/gov._mcauliffe_announces_virginias91759.asp.

39. Garrett Hardin, "The Tragedy of the Commons," *Science* 162, no. 3859 (1968): 1243–48.

40. Enrica Papa and António Ferreira, "'Moving to Access' in Transport Planning: Identifying Barriers, Designing Strategies" (presented at AESOP Congress, Lisbon, July 2017), http://westminsterresearch.wmin.ac.uk/19955/1/AESOP%20Enrica%20Papa%20Antonio%20Ferreira.pdf.

41. Transportation Research Board, *Highway Capacity Manual* (2016).

42. Ian Lockwood, "ITE Traffic Calming Definition," *ITE Journal* 67, no. 7 (1997): 22–24; Institute of Transportation Engineers website on Traffic Calming, http://www.ite.org/traffic/.

43. Boisjoly and Ahmed El-Geneidy, "Insider," 33–43.

44. Andres Duany and Emily Talen, "Making the Good Easy: The Smart Code Alternative," *Fordham Urban Law Journal* 29 (2001): 1445–68.

45. International Transport Forum, *Linking People and Places: New Ways of Understanding Spatial Access in Cities* (Paris: International Transport Forum, 2017), https://www.itf-oecd.org/linking-people-and-places; Scott Campbell, "Green Cities, Growing Cities, Just Cities?: Urban Planning and the Contradictions of Sustainable Development," *Journal of the American Planning Association* 62, no. 3 (1996): 296–312.

46. Michiel de Bok and Frank van Oort, "Agglomeration Economies, Accessibility and the Spatial Choice Behavior of Relocating Firms, *Journal of Transport and Land Use* 4, no. 1 (2011): 5–24.

APPENDIX A

1. Levine, Merlin, and Grengs. "Project-Level Accessibility Analysis," 107–19.

2. Grengs, et al., "Intermetropolitan Comparison," 427–43.

3. Schroeder et al., *Manual of Transportation Engineering Studies*.

4. For details on the described procedure, see Levine, Merlin, and Grengs, "Project-Level Accessibility Analysis," 107–19.

5. Nelson, *Planner's Estimating Guide*.

6. Federal Highway Administration, *Highway Statistics 2008*.

7. Levine et al., "Does Accessibility Require Density or Speed?," 157–72.

Bibliography

Accessibility Observatory, Center for Transportation Studies, University of Minnesota. Access Across America: Transit 2014 Data, 2014. http://access.umn.edu/data/datasets/.

Alonso, William. "A Theory of the Urban Land Market." In *Readings in Urban Economics*, edited by Matthew Edel and Jerome Rothenberg, 104–110. New York: Macmillan, 1972.

American Association of State Highway and Transportation Officials and Knovel. *A Policy on Geometric Design of Highways and Streets*. 6th. ed. Washington, DC: American Association of State Highway and Transportation Officials, 2011.

Apparicio, Philippe, and Anne-Marie Séguin. "Measuring the Accessibility of Services and Facilities for Residents of Public Housing in Montreal." *Urban Studies* 43, no. 1 (2006): 187–211.

ARC. *LCI Transportation Program*. Atlanta, GA: Atlanta Regional Commission, 2010. http://www.atlantaregional.com/land-use/livable-centers-initiative/lci-transportation-program.

Arlington County. *Additional Performance Measures*. Arlington County, Virginia, 2015. https://transportation.arlingtonva.us/wp-content/uploads/sites/19/2015/06/FY14_Transportation_Performance_Measures_Appendix.pdf.

Auld, Joshua, and Abolfazl Mohammadian. "Planning-Constrained Destination Choice in Activity-Based Model: Agent-Based Dynamic Activity Planning and Travel Scheduling." *Transportation Research Record: Journal of the Transportation Research Board*, no. 2254 (2011): 170–79.

Axhausen, Kay W. "Accessibility Long Term Perspectives." *Journal of Transport and Land Use* 1, no. 2 (2008): 5–22.

Bae, Chang-Hee Christine. "Air Quality and Travel Behavior: Untying the Knot." *Journal of the American Planning Association* 59, no. 1 (1993): 65–74.

Banister, David. "The Sustainable Mobility Paradigm." *Transport Policy* 15, no. 2 (2008): 73–80.

Banister, David, and Mark Thurstain-Goodwin. "Quantification of the Non-Transport Benefits Resulting from Rail Investment." *Journal of Transport Geography* 19, no. 2 (2011): 212–23.

Bannock Planning Organization. "Traffic Impact Study Guidelines." Produced for the City of Pocatello, City of Chubbuck, and Bannock County (Idaho), 2006. http://bannockplanning.org/wp-content/uploads/trafficimpactstudy.pdf.

Bartholomew, Keith, Reid Ewing, and Gail Meakins. "Integrated Transportation Scenario Planning." *Environment and Planning* FHWA-HEP-10-034 (July 2010): 1–229.

Baum-Snow, Nathaniel. "Did Highways Cause Suburbanization?" *Quarterly Journal of Economics* 122, no. 2 (2007): 775–805.

Ben-Akiva, Moshe E., and Steven E. Lerman. *Discrete Choice Analysis: Theory and Application to Predict Travel Demand*. Vol. 9. Cambridge, MA: MIT Press, 1985.

Benenson, Itzhak, Karel Martens, Yodan Rofé, and Ariela Kwartler. "Public Transport versus Private Car: GIS-Based Estimation of Accessibility Applied to the Tel Aviv Metropolitan Area." *Annals of Regional Science* 47, no. 3 (2011): 499–515.

Bhat, Chandra, Susan Handy, Kara Kockelman, Hani Mahmassani, Qinglin Chen, and Lisa Weston. *Accessibility Measures: Formulation Considerations and Current Applications*. Research Report 4938-2. Austin: Center for Transportation Research, Bureau of Engineering Research, University of Texas, 2000.

Bhattacharjee, Sutapa, and Andrew R. Goetz. "Impact of Light Rail on Traffic Congestion in Denver." *Journal of Transport Geography* 22 (2012): 262–70.

Black, John, and Chatherine M. Conroy. "Accessibility Measures and the Social Evaluation of Urban Structure." *Environment and Planning A* 9 (1977): 1013–31.

Blumenberg, Evelyn, and Michael Manville. "Beyond the Spatial Mismatch: Welfare Recipients and Transportation Policy." *Journal of Planning Literature* 19, no. 2 (2004): 182–205.

Boarnet, Marlon, and Randall Crane. *Travel by Design: The Influence of Urban Form on Travel*. New York: Oxford University Press, 2001.

Boarnet, Marlon G., and Andrew F. Haughwout. "Do Highways Matter?: Evidence and Policy Implications of Highways' Influence on Metropolitan Development." Berkeley: University of California Transportation Center, 2000.

Boisjoly, Geneviève, and Ahmed M. El-Geneidy. "How to Get There?: A Critical Assessment of Accessibility Objectives and Indicators in Metropolitan Transportation Plans." *Transport Policy* 55 (2017): 38–50.

Boisjoly, Geneviève, and Ahmed M. El-Geneidy. "The Insider: A Planners' Perspective on Accessibility." *Journal of Transport Geography* 64 (2017): 33–43.

Boisjoly, Geneviève, and Ahmed El-Geneidy. *Measuring Performance: Accessibility Metrics in Metropolitan Regions around the World*. Washington, DC: Brookings, 2017.

Bonavia, Michael R. *The Economics of Transport*. London: Cambridge at the University Press, 1936.

Boston Region Metropolitan Planning Organization. "Transportation Equity." In *Long-Range Transportation Plan of the Boston Region Metropolitan Planning Organization*: 7-1–7-17. Boston, MA: Boston Region Metropolitan Planning Organization, 2015.

Bullard, Robert D., Angel O. Torres, and Glenn S. Johnson, eds. *Sprawl City: Race, Politics, and Planning in Atlanta*. Washington, DC: Island Press, 2000.

Cairns, Shannon, Jessica Greig, and Martin Wachs. *Environmental Justice and Transportation: A Citizen's Handbook*. UCTC No. 620. Berkeley: University of California Transportation Center, 2003.

CalTrans. *Guide for the Preparation of Traffic Impact Studies*. Sacramento: Department of Transportation, State of California, December 2002.

CalTrans. *Performance Measure for Rural Transportation Systems: Guidebook* (2006), secs. 3-1 and 4-1. http://www.dot.ca.gov/perf/library/pdf/RSPMGuidebook.pdf.

Cambridge Systematics. *Technical Methods to Support Analysis of Environmental Justice Issues*. Cambridge, MA: National Cooperative Highway Research Program, 2002.

Campbell, Scott. "Green Cities, Growing Cities, Just Cities?: Urban Planning and the Contradictions of Sustainable Development." *Journal of the American Planning Association* 62, no. 3 (1996): 296–312.

Casas, Irene. "Evaluating the Importance of Accessibility to Congestion Response Using a GIS-Based Travel Simulator." *Journal of Geographical Systems* 5, no. 1 (2003): 109–27.

Cervero, Robert. "Bus Rapid Transit (BRT): An Efficient and Competitive Mode of Public Transport." Working Paper 2013-01. Berkeley: Institute of Urban and Regional Development, University of California. http://iurd.berkeley.edu/wp/2013-01.pdf.

Cervero, Robert. "Jobs-Housing Balancing and Regional Mobility." *Journal of the American Planning Association* 55, no. 2 (1989): 136–50.

Cervero, Robert. *Paradigm Shift: From Automobility to Accessibility Planning*. Working Paper 677. Berkeley: Institute of Urban and Regional Development, University of California, 1996.

Cervero, Robert. *The Transit Metropolis: A Global Inquiry*. Washington, DC: Island Press, 1998.

Cervero, Robert, and Jin Murakami. "Effects of Built Environments on Vehicle Miles Traveled: Evidence from 370 US Urbanized Areas." *Environment and Planning A* 42, no. 2 (2010): 400–418.

Chatman, Daniel G. "Residential Choice, the Built Environment, and Nonwork Travel: Evidence Using New Data and Methods." *Environment and Planning A* 41, no. 5 (2009): 1072–89.

Chen, Cynthia. *The Temporal and Social Dimension of Accessibility for New York City Residents*. RF 49777-13-18. New York: University Transportation Research Center, City College of New York, 2008.

Chen, Jason, Cynthia Chen, and Harry Timmermans. "Accessibility Trade-Offs in Household Residential Location Decisions." *Transportation Research Record: Journal of the Transportation Research Board*, no. 2077 (2008): 71–79.

Cheng, Jianquan Q., Luca Bertolini, and Frank le Clercq. "Measuring Sustainable Accessibility." *Transportation Research Record: Journal of the Transportation Research Board*, no. 2017 (2007): 16–25.

Cheshire, Paul, and Stephen Sheppard. "On the Price of Land and the Value of Amenities." *Economica* 62, no. 246 (1995): 247–67.

Chicago Metropolitan Agency for Planning. *GO TO 2040: Comprehensive Regional Plan*. Chicago: Chicago Metropolitan Agency for Planning, 2010.

Chicago Metropolitan Agency for Planning. *Preferred Regional Scenario*. Chicago: Chicago Metropolitan Agency for Planning, 2008.

Claritas, Inc. *Claritas Business-Facts Methodology*. Ithaca, NY: Claritas, 2002.

Clinton, William J. *Executive Order No. 12898: Federal Actions to Address Environmental Justice in Minority Populations and Low-Income Populations. Federal Register* 59:7629 (1994).

Cournot, Augustin. *Researches into the Mathematical Principles of the Theory of Wealth*. Translated by Nathaniel T. Bacon with a Bibliography of Mathematical Economics by Irving Fisher. 1838. Reprint, New York: MacMillan, 1897.

Crane, Randall. "Counterpoint: Accessibility and Sprawl." *Journal of Transport and Land Use* 1, no. 1 (2008): 13–19.

Crane, Randall. "On Form versus Function: Will the New Urbanism Reduce Traffic, or Increase It?" *Journal of Planning Education and Research* 15, no. 2 (1996): 117–26.

Currie, Graham. "Bus Transit Oriented Development—Strengths and Challenges Relative to Rail." *Journal of Public Transportation* 9, no. 4 (2006): 1–21.

Dalumpines, Ron, and Darren M. Scott. "Determinants of Route Choice Behavior: A Comparison of Shop versus Work Trips Using the Potential Path Area-Gateway (PPAG) Algorithm and Path-Size Logit." *Journal of Transport Geography* 59 (2017): 59–68.

de Bok, Michiel, and Frank van Oort. "Agglomeration Economies, Accessibility and the Spatial Choice Behavior of Relocating Firms. *Journal of Transport and Land Use* 4, no. 1 (2011): 5–24.

de Souza Briggs, Xavier, ed. *The Geography of Opportunity: Race and Housing Choice in Metropolitan America*. Washington, DC: Brookings Institution, 2005.

Deakin, Elizabeth. "Equity and Environmental Justice in Sustainable Transportation: Toward a Research Agenda." In *Institutions and Sustainable Transport: Regulatory Reform in Advanced Economies*, edited by Piet Rietveld and Roger R. Stough, 51–69. Cheltenham, UK: Edward Elgar, 2007.

Deka, Devajyoti. "Social and Environmental Justice Issues in Urban Transportation." In *The Geography of Urban Transportation*, edited by Susan Hanson and Genevieve Giuliano, 332–55. 3rd ed. New York: Guilford Press, 2004.

Dodson, Jago, Matthew Burke, Rick Evans, Brendan Gleeson, and Neil Sipe. "Travel Behavior Patterns of Different Socially Disadvantaged Groups: Analysis of Household Travel Survey Data for a Dispersed Metropolitan Area." *Transportation Research Record: Journal of the Transportation Research Board*, no. 2163 (2010): 24–31.

Dreier, Peter, John H. Mollenkopf, and Todd Swanstrom. *Place Matters: Metropolitics for the Twenty-First Century*. 2nd ed. Lawrence: University Press of Kansas, 2004.

Du, Hongbo, and Corinne Mulley. "Relationship between Transport Accessibility and Land Value: Local Model Approach with Geographically Weighted Regression." *Transportation Research Record: Journal of the Transportation Research Board*, no. 1977 (2006): 197–205.

Duany, Andres, and Emily Talen. "Making the Good Easy: The Smart Code Alternative." *Fordham Urban Law Journal* 29 (2001): 1445–68.

Dupuit, Jules. "On the Measurement of the Utility of Public Works." In *Readings in Welfare Economics*, by Kenneth J. Arrow and Tibor Scitovsky. Homewood, IL: Richard D. Irwin, 1969. Selected by a committee of the American Economic Association. Translated in *International Economic Papers* 2 (1952): 83–110. English translation by R. H. Barback from "De la Mesure de l'Utilité des Travaux Publics," *Annales des Ponts et Chausées*, 2nd ser., vol. 8, 1844.

Duranton, Gilles, and Erick Guerra. *Developing a Common Narrative on Urban Accessibility: An Urban Planning Perspective*. Washington DC: Brookings, 2016.

Echenique, Marcial H., Anthony J. Hargreaves, Gordon Mitchell, and Anil Namdeo. "Does Urban Form Really Matter?" *Journal of the American Planning Association* 78, no. 2 (2012): 121–37.

Edwards, John D., Jr. *Transportation Planning Handbook*. Englewood Cliffs, NJ: Prentice-Hall, 1992.

Envision Utah. *Wasatch Choices 2040: A Four County Land-Use and Transportation Vision*. Salt Lake City, UT: Envision Utah, 2006.

Ermagun, Alireza, and David Levinson. "Accessibility and Transit Performance." Minneapolis: University of Minnesota Nexus Working Papers 000129 (2015). https://conservancy.umn.edu/bitstream/handle/11299/179832/Accessibility-TransitPerformance.pdf?sequence=1&isAllowed=y.

Ettema, Dick, and Harry Timmermans. "Space-Time Accessibility under Conditions of Uncertain Travel Times: Theory and Numerical Simulations." *Geographical Analysis* 39, no. 2 (2007): 217–40.

Ewing, Reid. "Characteristics, Causes, and Effects of Sprawl: A Literature Review." *Urban Ecology* 21, no. 2 (1994): 519–35.

Ewing, Reid. "Measuring Transportation Performance." *Transportation Quarterly* 49, no. 1 (1995): 91–104.

Ewing, Reid, and Robert Cervero. "Travel and the Built Environment: A Meta-Analysis." *Journal of the American Planning Association* 76, no. 3 (2010): 265–94.

Ewing, Reid, Shima Hamidi, Guang Tian, David Proffitt, Stefania Tonin, and Laura Fregolent. "Testing Newman and Kenworthy's Theory of Density and Automobile Dependence." *Journal of Planning Education and Research* 38, no. 2 (2017): 167–82.

Ewing, Reid, Rolf Pendall, and Don Chen. *Sprawl Scores for 83 Metropolitan Regions*. Washington, DC: Smart Growth America, 2002.

Fan, Yingling. "Household Structure and Gender Differences in Travel Time: Spouse/Partner Presence, Parenthood, and Breadwinner Status." *Transportation* 44, no. 2 (2017): 271–91.

Fan, Yingling, Andrew E. Guthrie, and David M. Levinson. "Impact of Light Rail Implementation on Labor Market Accessibility: A Transportation Equity Perspective." *Journal of Transport and Land Use* 5, no. 3 (2012): 28–39.

Farber, Steven, Melinda Z. Morang, and Michael J. Widener. "Temporal Variability in Transit-Based Accessibility to Supermarkets." *Applied Geography* 53 (2014): 149–59.

Farrington, John, and Conor Farrington. "Rural Accessibility, Social Inclusion and Social Justice: Towards Conceptualisation." *Journal of Transport Geography* 13, no. 1 (2005): 1–12.

Federal Highway Administration. *Highway Functional Classification Concepts, Criteria and Procedures* (U.S. Department of Transportation). Accessed September 19, 2016. https://www.fhwa.dot.gov/planning/processes/statewide/related/highway_functional_classifications/section02.cfm.

Federal Highway Administration. *Highway Statistics 2008*. Washington, DC: Bureau of Highway Statistics, 2008. https://www.fhwa.dot.gov/policyinformation/statistics/2008/vm1.cfm.

Federal Transit Administration. *FTA Circular 4702.1A: Title VI and Title VI–Dependent Guidelines for Federal Transit Administration Recipients*. Washington, DC: U.S. Department of Transportation, 2007.

Federal Transit Administration. *FTA Circular 4702.1B: Title VI Requirements and Guidelines for Federal Transit Administration Recipients*. Washington, DC: U.S. Department of Transportation, 2012.

Ferguson, Erin M., Jennifer Duthie, Avinash Unnikrishnan, and S. Travis Waller. "Incorporating Equity into the Transit Frequency-Setting Problem." *Transportation Research Part A* 46, no. 1 (2012): 190–99.

Flyvbjerg, Bent. *Making Social Science Matter: Why Social Inquiry Fails and How It Can Succeed Again*. New York: Cambridge University Press. 2001.

Forkenbrock, David J., and Lisa A. Schweitzer. "Environmental Justice in Transportation Planning." *Journal of the American Planning Association* 65, no. 1 (1999): 96–111.

Forkenbrock, David J., and Glen E. Weisbrod. *Guidebook for Estimating the Social and Economic Effects of Transportation Projects*. NCHRP Report 456. Washington, DC: National Academy Press, 2001.

Foth, Nicole, Kevin Manaugh, and Ahmed M. El-Geneidy. "Towards Equitable Transit: Examining Transit Accessibility and Social Need in Toronto, Canada, 1996–2006." *Journal of Transport Geography* 29 (2013): 1–10.

Galster, George, Royce Hanson, Michael R. Ratcliffe, Harold Wolman, Stephen Coleman, and Jason Freihage. "Wrestling Sprawl to the Ground: Defining and Measuring an Elusive Concept." *Housing Policy Debate* 12, no. 4 (2001): 681–717.

Garrett, Mark, and Brian Taylor. "Reconsidering Social Equity in Public Transit." *Berkeley Planning Journal* 13 (1999): 6–27.

Geurs, Karst T., Michiel de Bok, and Barry Zondag. "Accessibility Benefits of Integrating Land Use and Public Transport Policy Plans in the Netherlands." In *Accessibility Analysis and Transport Planning: Challenges for Europe and North America*, edited by Karst T. Geurs, Kevin J. Krizek, and Aura Reggiani, 135–53. Cheltenham, UK: Edward Elgar, 2012.

Geurs, Karst T., and Bert van Wee. "Accessibility Evaluation of Land-Use and Transport Strategies: Review and Research Directions." *Journal of Transport Geography* 12, no. 2 (2004): 127–40.

Ghosh, Avijit, and C. Samuel Craig. "Formulating Retail Location Strategy in a Changing Environment." *Journal of Marketing* 47, no. 3 (1983): 53–68.

Giuliano, Genevieve. "Travel, Location and Race/Ethnicity." *Transportation Research Part A* 37, no. 4 (2003): 351–72.

Gjestland, Arnstein, David McArthur, Liv Osland, and Inge Thorsen. "A Bridge over Troubled Waters: Valuing Accessibility Effects of a New Bridge." In *Accessibility Analysis and Transport Planning: Challenges for Europe and North America*, edited by Karst T. Geurs, Kevin J. Krizek, and Aura Reggiani, 173–92. Cheltenham, UK: Edward Elgar, 2012.

Glaeser, Edward L., Matthew E. Kahn, and Jordan Rappaport. "Why Do the Poor Live in Cities: The Role of Public Transportation." *Journal of Urban Economics* 63, no. 1 (2008): 1–24.

Gobillon, Laurent, Harris Selod, and Yves Zenou. "The Mechanisms of Spatial Mismatch." *Urban Studies* 44, no. 12 (2007): 2401–27.

Golub, Aaron, Richard A. Marcantonio, and Thomas W. Sanchez. "Race, Space, and Struggles for Mobility." *Urban Geography* 34, no. 5 (2013): 699–728.

Golub, Aaron, Glenn Robinson, and Brendan Nee. "Making Accessibility Analyses Accessible: A Tool to Facilitate the Public Review of the Effects of Regional Transportation Plans on Accessibility." *Journal of Transport and Land Use* 6, no. 3 (2013): 17–28.

Gong, Jianxin. "Clarifying the Standard Deviational Ellipse." *Geographical Analysis* 34, no. 2 (2002): 155–67.

Gordon, Peter, Ajay Kumar, and Harry W. Richardson. "Beyond the Journey to Work." *Transportation Research Part A* 22, no. 6 (1988): 419–26.

Governor's Office of Planning and Research (CA). "Revised Proposal on Updates to the CEQA Guidelines on Evaluating Transportation Impacts in CEQA." Sacramento, CA, 2016. https://www.opr.ca.gov/docs/Revised_VMT_CEQA_Guidelines_Proposal_January_20_2016.pdf.

Grengs, Joe. "Community-Based Planning as a Source of Political Change: The Transit Equity Movement of Los Angeles' Bus Riders Union." *Journal of the American Planning Association* 68, no. 2 (2002): 165–78.

Grengs, Joe. "Comparing People and Places with Transportation Accessibility in Metropolitan Detroit." In *Mapping Detroit: Land, Community, and Shaping a City*, edited by June M. Thomas and Henco Bekkering, 101–14. Detroit, MI: Wayne State University Press, 2015.

Grengs, Joe. "Equity and the Social Distribution of Job Accessibility in Detroit." *Environment and Planning B* 39, no. 5 (2012): 785–800.

Grengs, Joe. "Job Accessibility and the Modal Mismatch in Detroit." *Journal of Transport Geography* 18, no. 1 (2010): 42–54.

Grengs, Joe. "Measuring Change in Small-Scale Transit Accessibility with Geographic Information Systems: The Cases of Buffalo and Rochester." *Transportation Research Record: Journal of the Transportation Research Board*, no. 1887 (2004): 10–17.

Grengs, Joe. "Nonwork Accessibility as a Social Equity Indicator." *International Journal of Sustainable Transportation* 9, no. 1 (2015): 1–14.

Grengs. Joe, Jonathan Levine, Qing Shen, and Qingyun Shen. "Intermetropolitan Comparison of Transportation Accessibility: Sorting Out Mobility and Proximity in San Francisco and Washington, DC." *Journal of Planning Education and Research* 29, no. 4 (2010): 427–43.

Guerra, Erick. "Valuing Rail Transit: Comparing Capital and Operating Costs with Consumer Benefits." *Transportation Research Record: Journal of the Transportation Research Board*, no. 2219 (2011): 50–58.

"Guidelines for Traffic Impact Analysis (Zoning Petitions)." Wake County, North Carolina. Accessed February 1, 2019. http://www.wakegov.com/planning/zoning/Documents/guidelines_for_traffic_impact_analysisUDO3.pdf.

Gulhan, Gorkem, Huseyin Ceylan, Ozgur Baskan, and Halim Ceylan. "Using Potential Accessibility Measure for Urban Public Transportation Planning: A Case Study of Denizli, Turkey." *PROMET–Traffic & Transportation* 26, no. 2 (2014): 129–37.

Gutman, Jeffrey, and Adie Tomer. *Developing a Common Narrative on Urban Accessibility: Overview*. Moving to Access. Washington, DC: Brookings, 2016. https://www.brookings.edu/wp-content/uploads/2017/01/overview-digital.pdf.

Habib, Muhammad A., and Eric J. Miller. "Reference-Dependent Residential Location Choice Model within a Relocation Context." *Transportation Research Record: Journal of the Transportation Research Board*, no. 2133 (2009): 92–99.

Hägerstrand, Torsten. "What About People in Regional Science?" *Papers in Regional Science* 24, no. 1 (1970): 6–21.

Haig, Robert M. "The Assignment of Activities to Areas in Urban Regions." In *Regional Survey of New York and Its Environs*, 1:31–44. New York: Regional Survey, 1927. Originally published in *Quarterly Journal of Economics* 40 (1926): 402–34.

Haig, Robert M. *Major Economic Factors in Metropolitan Growth and Its Arrangement: A Study of Trends and Tendencies in the Economic Activities within the Region of New York and Its Environs*. New York: Committee on Regional Plan of New York and Its Environs, 1927.

Halden, Derek. "The Use and Abuse of Accessibility Measures in UK Passenger Transport Planning." *Research in Transportation Business & Management* 2 (2011): 12–19.

Halden, Derek. "Using Accessibility Measures to Integrate Land Use and Transport Policy in Edinburgh and the Lothians." *Transport Policy* 9, no. 4 (2002): 313–24.

Hamidi, Shima, Reid Ewing, Ilana Preuss, and Alex Dodds. "Measuring Sprawl and Its Impacts: An Update." *Journal of Planning Education and Research* 35, no. 1 (2015): 35–50.

Handy, Susan L. *Accessibility- vs. Mobility-Enhancing Strategies for Addressing Automobile Dependence in the U.S.* UCD-ITS-RR-02-15. Davis: Institute of Transportation Studies, University of California, 2002.

Handy, Susan L., and Deb A. Niemeier. "Measuring Accessibility: An Exploration of Issues and Alternatives." *Environment and Planning A* 29, no. 7 (1997): 1175–94.

Hansen, Walter G. "How Accessibility Shapes Land Use." *Journal of the American Planning Association* 25, no. 2 (1959): 73–76.

Hansen, Willard B. "An Approach to the Analysis of Metropolitan Residential Extension." *Journal of Regional Science* 3, no. 1 (1961): 37–55.

Hardin, Garrett. "The Tragedy of the Commons." *Science* 162, no. 3859 (1968): 1243–48.

Hatzopoulou, Marianne, and Eric J. Miller. "Institutional Integration for Sustainable Transportation Policy in Canada." *Transport Policy* 15, no. 3 (2008): 149–62.

Hau, Timothy D. *Economic Fundamentals of Road Pricing*. Washington, DC: Infrastructure and Urban Development Department, World Bank, WPS 1070, 1992.

Hay, Alan. "Equity and Welfare in the Geography of Public Transport Provision." *Journal of Transport Geography* 1, no. 2 (1993): 95–101.

Hayes, M. Cordey, and Alan G. Wilson. "Spatial Interaction." *Socio-Economic Planning Sciences* 5, no. 1 (1971): 73–95.

Heilig, Michael, Nicolai Mallig, Tim Hilgert, Martin Kagerbauer, and Peter Vortisch. "Large-Scale Application of a Combined Destination and Mode Choice Model Estimated with Mixed Stated and Revealed Preference Data." *Transportation Research Record: Journal of the Transportation Research Board*, no. 2669 (2017): 31–40.

Hensher, David A. "Measurement of the Valuation of Travel Time Savings." *Journal of Transport Economics and Policy* 35, no. 1 (2001): 71–98.

Hess, Daniel B. "Access to Employment for Adults in Poverty in the Buffalo-Niagara Region." *Urban Studies* 42, no. 7 (2005): 1177–200.

Hess, Daniel B., Jeffrey Brown, and Donald Shoup. "Waiting for the Bus." *Journal of Public Transportation* 7, no. 4 (2004): 67–84.

Hess, Daniel B., Brian Taylor, and Allison Yoh. "Light Rail Lite or Cost-Effective Improvements to Bus Service?: Evaluating Costs of Implementing Bus Rapid Transit." *Transportation Research Record: Journal of the Transportation Research Board*, no. 1927 (2005): 22–30.

Higgins, Christopher, Mark Ferguson, and Pavlos Kanaroglou. "Light Rail and Land Use Change: Rail Transit's Role in Reshaping and Revitalizing Cities." *Journal of Public Transportation* 17, no. 2 (2014): 93–112.

Hill, Donald M. "A Growth Allocation Model for the Boston Region." *Journal of the American Institute of Planners* 31, no. 2 (1965): 111–20.

Hodge, David C. "My Fair Share: Equity Issues in Urban Transportation." In *The Geography of Urban Transportation*, edited by Susan Hanson, 359–75. 2nd ed. New York: Guilford Press, 1995.

Holzer, Harry. "The Spatial Mismatch Hypothesis: What Has the Evidence Shown?" *Urban Studies* 28, no. 1 (1991): 105–22.

Horner, Mark W. "Exploring Metropolitan Accessibility and Urban Structure." *Urban Geography* 25, no. 3 (2004): 264–84.

Houseman, Gerald L. *The Right of Mobility*. Port Washington, NY: Kennikat Press, 1979.

Hu, Lingqian. "Job Accessibility of the Poor in Los Angeles." *Journal of the American Planning Association* 81, no. 1 (2015): 30–45.

Hupkes, Geurt. "The Law of Constant Travel Time and Trip-Rates." *Futures* 14, no. 1 (1982): 38–46.

Hurd, Richard M. *Principles of City Land Values*. New York: Record and Guide, 1903.

Ingram, David R. "The Concept of Accessibility: A Search for an Operational Form." *Regional Studies* 5, no. 2 (1971): 101–7.

Innes, Judith E. "Information in Communicative Planning." *Journal of the American Planning Association* 68 (1998): 52–63.

Institute for Transportation and Development Policy. *The Bus Rapid Transit Standard*. New York, 2016. https://www.itdp.org/library/standards-and-guides/the-bus-rapid-transit-standard/the-scorecard/.

Institute of Transportation Engineers. Website on Traffic Calming. http://www.ite.org/traffic/.

Institute of Transportation Engineers. *Traffic Engineering Handbook*. Hoboken, NJ: Wiley, 2016.

Institute of Transportation Engineers. *Transportation Impact Analyses for Site Development: An ITE Recommended Practice*. Washington, DC, 2010.

International Transport Forum. *Linking People and Places: New Ways of Understanding Spatial Access in Cities*. Paris: International Transport Forum, 2017. https://www.itf-oecd.org/linking-people-and-places.

Isard, Walter, *Methods of Regional Analysis: An Introduction to Regional Science*. Cambridge, MA: MIT Press, 1960.

Jaffe, Eric. "Explaining Transit's Secret Language." *CityLab* (March 8, 2012). https://www.citylab.com/transportation/2012/03/teaching-language-transit/1438/.

Jain, Juliet, and Glen Lyons. "The Gift of Travel Time." *Journal of Transport Geography* 16, no. 2 (2008): 81–89.

Johnston-Anumonwo, Ibipo. "Racial Differences in the Commuting Behavior of Women in Buffalo, 1980–1990." *Urban Geography* 16, no. 1 (1995): 23–45.

Kain, John F. "The Spatial Mismatch Hypothesis: Three Decades Later." *Housing Policy Debate* 3, no. 2 (1992): 371–460.

Karner, Alex, and Deb A. Niemeier. "Civil Rights Guidance and Equity Analysis Methods for Regional Transportation Plans: A Critical Review of Literature and Practice." *Journal of Transport Geography* 33 (2013): 126–34.

Karsch v. Camden County Board of Adjustments. 302 S.W.3d 754 (M.O. App. 2010), 11.

Kawabata, Mizuki. "Spatiotemporal Dimensions of Modal Accessibility Disparity in Boston and San Francisco." *Environment and Planning A* 41, no. 1 (2009): 183–98.

Kawabata, Mizuki, and Qing Shen. "Commuting Inequality between Cars and Public Transit: The Case of the San Francisco Bay Area, 1990–2000." *Urban Studies* 44, no. 9 (2007): 1759–80.

Kawabata, Mizuki, and Qing Shen. "Job Accessibility as an Indicator of Auto-Oriented Urban Structure: A Comparison of Boston and Los Angeles with Tokyo." *Environment and Planning B* 33, no. 1 (2006): 115–30.

Krause, Andy L., and Christopher Bitter. "Spatial Econometrics, Land Values, and Sustainability: Trends in Real Estate Valuation Research." Supplement, *Cities* 29, no. 2 (2012): S19–25.

Krizek, Kevin J. "Operationalizing Neighborhood Accessibility for Land Use–Travel Behavior Research and Regional Modeling." *Journal of Planning Education and Research* 22, no. 3 (2003): 270–87.

Kwan, Mei-Po. "Space-Time and Integral Measures of Individual Accessibility: A Comparative Analysis Using a Point-Based Framework." *Geographical Analysis* 30, no. 3 (1998): 191–216.

Kwan, Mei-Po, and Joe Weber. "Scale and Accessibility: Implications for the Analysis of Land Use–Travel Interaction." *Applied Geography* 28, no. 2 (2008): 110–23.

Kwok, Rebecca C. W., and Anthony G. O. Yeh. "The Use of Modal Accessibility Gap as an Indicator for Sustainable Transport Development." *Environment and Planning A* 36 (2004): 921–36.

Lansing, John B., Robert W. Marans, and Robert B. Zehner. *Planned Residential Environments*. Ann Arbor, MI: Survey Research Center, Institute for Social Research, 1970.

Leathers, Nancy J. "Residential Location and Mode of Transportation to Work: A Model of Choice." *Transportation Research* 1, no. 2 (1967): 129–55.

Lee, Brian H. Y., Paul Waddell, Liming Wang, and Ram M. Pendyala. "Reexamining the Influence of Work and Nonwork Accessibility on Residential Location Choices with a Microanalytic Framework." *Environment and Planning A* 42, no. 4 (2010): 913–30.

Lee, Bumsoo. "'Edge' or 'Edgeless' Cities?: Urban Spatial Structure in US Metropolitan Areas, 1980 to 2000." *Journal of Regional Science* 47, no. 3 (2007): 479–515.

Levine, Jonathan. "Is Bus versus Rail Investment a Zero-Sum Game?: The Misuse of the Opportunity-Cost Concept." *Journal of the American Planning Association* 79, no. 1 (2013): 5–15.

Levine, Jonathan. "Urban Transportation and Social Equity: Transportation-Planning Paradigms That Impede Policy Reform." In *Policy, Planning, and People: Promoting Justice in Urban Development*, edited by Naomi Carmon and Susan S. Fainstein, 141–60. Philadelphia: University of Pennsylvania Press, 2013.

Levine, Jonathan. *Zoned Out: Regulations, Market, and Choices in Transportation and Metropolitan Land Use.* Washington, DC: Resources for the Future, 2006.

Levine, Jonathan, Joe Grengs, Qingyun Shen, and Qing Shen. "Does Accessibility Require Density or Speed?: A Comparison of Fast versus Close in Getting Where You Want to Go in U.S. Metropolitan Regions." *Journal of the American Planning Association* 78, no. 2 (2012): 157–72.

Levine, Jonathan, Joe Grengs, Qingyun Shen, and Qing Shen, *Metropolitan Accessibility and Transportation Sustainability: Comparative Indicators for Policy Reform.* Final Report from EPA Agreement Number: RD-833334901-0 and FHWA Cooperative Agreement Number: DTFH61-07-H-00037, January 2011. http://hdl.handle.net/2027.42/147459.

Levine, Jonathan, Louis Merlin, and Joe Grengs. "Project-Level Accessibility Analysis for Land-Use Planning." *Transport Policy* 53 (2017): 107–19. https://doi.org/10.1016/j.tranpol.2016.09.005.

Levinson, David, Bernadette Marion, and Michael Iacono. *Access to Destinations, Phase 3: Measuring Accessibility by Automobile.* Mn/DOT 2010-09. Minneapolis: Department of Civil Engineering, University of Minnesota, 2010.

Levinson, David M., Wes Marshall, and Kay Axhausen. *Elements of Access: Transport Planning for Engineers, Transport Engineering for Planners.* Sydney: Network Design Laboratory, 2017.

Lindblom, Charles E. "The Science of 'Muddling Through.'" *Public Administration Review* 19, no. 2 (1959): 79–88.

Litman, Todd. "Evaluating Transportation Equity." *World Transport Policy and Practice* 8, no. 2 (2002): 50–65.

Lockwood, Ian. "ITE Traffic Calming Definition." *ITE Journal* 67, no. 7 (1997): 22–24.

Lovelady, Adam. "MPOs and the Integration of Transportation and Land Use Planning." *Virginia Environmental Law Journal* 27, no. 3 (2009): 275–320.

Lowe, Kate, and Joe Grengs. "Private Donations for Public Transit: The Equity Implications of Detroit's Public-Private Streetcar." *Journal of Planning Education and Research* (March 13, 2018). https://doi.org/10.1177/0739456X18761237.

Lucas, Karen. "A Critical Assessment of Accessibility Planning for Social Inclusion." In *Accessibility Analysis and Transport Planning: Challenges for Europe and North America*, edited by Karst T. Geurs, Kevin J. Krizek, and Aura Reggiani, 228–42. Cheltenham, UK: Edward Elgar, 2012.

Lucas, Karen. "Transport and Social Exclusion: Where Are We Now?" *Transport Policy* 20 (2012): 105–13.

Luo, Wei, and Fahui Yang. "Measures of Spatial Accessibility to Healthcare in a GIS Environment: Synthesis and a Case Study in the Chicago Region." *Environment and Planning B*, no. 3 (2003): 865–84.

Ma, Kang-Rae, and David Banister. "Excess Commuting: A Critical Review." *Transport Reviews* 26, no. 6 (2006): 749–67.

MacKinnon, Ross D., and M. John Hodgson. "Optimal Transportation Networks: A Case Study of Highway Systems." *Environment and Planning* 2, no. 3 (1970): 267–84.

Manaugh, Kevin, Madhav G. Badami, and Ahmed M. El-Geneidy. "Integrating Social Equity into Urban Transportation Planning: A Critical Evaluation of Equity Objectives and Measures in Transportation Plans in North America." *Transport Policy* 37 (2015): 167–76.

Mann, Eric. "Confronting Transit Racism in Los Angeles." In *Just Transportation: Dismantling Race and Class Barriers to Mobility*, edited by Robert D. Bullard and Glenn S. Johnson, 68–83. Gabriola Island, BC: New Society, 1997.

Marshall, Alfred, *Principles of Economics*. New York: Macmillan, 1895.

Martens, Karel. "Justice in Transport as Justice in Accessibility: Applying Walzer's 'Spheres of Justice' to the Transport Sector." *Transportation* 39, no. 6 (2012): 1035–53.

Martens, Karel. *Transport Justice: Designing Fair Transportation Systems*. New York: Routledge, 2017.

Martens, Karel, Aaron Golub, and Glenn Robinson. "A Justice-Theoretic Approach to the Distribution of Transportation Benefits: Implications for Transportation Planning Practice in the United States." *Transportation Research Part A* 46, no. 4 (2012): 684–95.

Martin, Juan Carlos, and Aura Reggiani. "Recent Methodological Developments to Measure Spatial Interaction: Synthetic Accessibility Indices Applied to High-Speed Train Investments." *Transport Reviews* 27, no. 5 (2007): 551–71.

Massey, Douglas S., and Nancy A. Denton. *American Apartheid: Segregation and the Making of the Underclass*. Cambridge, MA: Harvard University Press, 1993.

McIntosh, James, Roman Trubka, Jeff Kenworthy, and Peter Newman. "The Role of Urban Form and Transit in City Car Dependence: Analysis of 26 Global Cities from 1960 to 2000." *Transportation Research Part D* 33 (2014): 95–110.

Merlin, Louis A. "Measuring Community Completeness: Jobs Housing Balance, Accessibility, and Convenient Local Access to Nonwork Destinations." *Environment and Planning B* 41, no. 4 (2014): 736–56. http://doi.org/10.1068/b120010p.

Merlin, Louis A. "A Portrait of Accessibility Change for Four US Metropolitan Areas." *Journal of Transport and Land Use* 10, no. 1 (2017): 309–36. http://dx.doi.org/10.5198/jtlu.2015.808.

Merlin, Louis A., and Lingqian Hu. "Does Competition Matter in Measures of Job Accessibility?: Explaining Employment in Los Angeles." Unpublished manuscript, 2017.

Merlin, Louis A., Jonathan Levine, and Joe Grengs. "Accessibility Analysis for Transportation Projects and Plans." *Transport Policy* 69 (2018): 35–48.

Metropolitan Transportation Commission. *Transportation 2035 Plan for the San Francisco Bay Area: Equity Analysis Report*. Oakland, CA: Metropolitan Transportation Commission, 2009.

Metropolitan Planning Commission and Association of Bay Area Governments. Equity Analysis Report: Plan Bay Area 2040 Final Supplemental Report (San Francisco: Metropolitan Transportation Commission and Association of Bay Area Governments, 2017). http://2040.planbayarea.org/sites/default/files/2017-07/Equity_Report_PBA%202040%20_7-2017.pdf.

Metropolitan Transportation Commission and Association of Bay Area Governments, *Bay Area Census*. http://www.bayareacensus.ca.gov/transportation/Commutetime19802000.htm.

Metz, David. "The Myth of Travel Time Saving." *Transport Reviews* 28, no. 3 (2008): 321–36.

Meyer, Jonas, Henrik Becker, Patrick M. Bosch, and Kay W. Axhausen."Autonomous Vehicles: The Next Jump in Accessibilities?" *Research in Transportation Economics* 62 (2017): 80–90.

Meyer, Michael D., and Eric J. Miller. *Urban Transportation Planning: A Decision-Oriented Approach*. 2nd ed. New York: McGraw-Hill, 2001.

Mid-America Regional Council. *Environmental Justice Analysis*. Kansas City, MO: Mid-America Regional Council, 2013.

Mid-Ohio Regional Planning Commission. "Appendix D: Environmental Justice Technical Analysis." In *2012–2035 Metropolitan Transportation Plan*. Columbus: Mid-Ohio Regional Planning Commission, 2012.

Mid-Region Council of Governments. *Economic Impact Report*. Albuquerque, NM, 2011.

Mid-Region Council of Governments and Mid-Region Metropolitan Planning Organization. "A Profile in Congestion (2012): The 30 Most Congested Corridors in the Albuquerque Metropolitan Planning Area." https://www.mrcog-nm.gov/images/stories/pdf/transportation/CMP/2012-cmp-profile.pdf.

Milakis, Dimitris, Robert Cervero, and Bert van Wee. "Stay Local or Go Regional?: Urban Form Effects on Vehicle Use at Different Spatial Scales: A Theoretical Concept and Its Application to the San Francisco Bay Area." *Journal of Transport and Land Use* 8, no. 2 (2015): 59–86.

Mill, John Stuart. *On Liberty*. 1859. Reprint, New York: Penguin Press, 1985.

Miller, Harvey J. "Measuring Space-Time Accessibility Benefits within Transportation Networks: Basic Theory and Computational Procedures." *Geographical Analysis* 31, no. 1 (1999): 1–26.

Mills, Scott G., and Sieglinde K. Neuhauser. "Quantitative Methods for Environmental Justice Assessment of Transportation." *Risk Analysis* 20, no. 3 (2000): 377–84.

Minnesota Population Center. *National Historical Geographic Information System*. Version 2.0. Minneapolis: University of Minnesota, 2011. http://www.nhgis.org.

Minocha, Inshu, P. S. Sriraj, Paul Metaxatos, and Piyushimita (Vonu) Thakuriah. "Analysis of Transit Quality of Service and Employment Accessibility for the Greater Chicago, Illinois, Region." *Transportation Research Record: Journal of the Transportation Research Board*, no. 2042 (2008): 20–29.

Mitchell, Robert B., and Chester Rapkin. *Urban Traffic: A Function of Land Use*. New York: Columbia University Press, 1954.

Mohammad, Sara I., Daniel J. Graham, Patricia C. Melo, and Richard J. Anderson. "A Meta-Analysis of the Impact of Rail Projects on Land and Property Values." *Transportation Research A* 50 (2013): 158–70.

Mok, Diana. "Do Two-Earner Households Base Their Choice of Residential Location on Both Incomes?" *Urban Studies* 44, no. 4 (2007): 723–50.

Mokhtarian, Patricia L., and Ilan Salomon. "How Derived Is the Demand for Travel?: Some Conceptual and Measurement Considerations." *Transportation Research Part A* 35, no. 8 (2001): 695–719.

Mokhtarian, Patricia L., and Ilan Salomon. "Travel for the Fun of It." *Access* 15 (1999): 26–31.

Mokhtarian, Patricia L., Ilan Salomon, and Lothlorien S. Redmond. "Understanding the Demand for Travel: It's Not Purely 'Derived.'" *Innovation: The European Journal of Social Science Research* 14, no. 4 (2001): 355–80.

Monschein, Andrew, and Brian D. Taylor. "Is Traffic Congestion Overrated?: Examining the Highly Variable Effects of Congestion on Travel and Accessibility." *Journal of Transport Geography* 64 (2017): 65–76.

Moore, George Edward. *Principia Ethica*. Edited by Thomas Baldwin. Cambridge: Cambridge University Press, 1993.

Moore, José, Karen Lucas, and John Bates. *Social Disadvantage and Transport in the UK: A Trip-Based Approach*. Oxford: Transport Studies Unit, University of Oxford, 2013.

Morlok, Edward K. *Introduction to Transportation Engineering and Planning*. New York: McGraw-Hill, 1978.

Morris, Jennifer M., Peter L. Dumble, and Marcus R. Wigan. "Accessibility Indicators for Transport Planning." *Transportation Research Part A* 13, no. 2 (1979): 91–109.

Murray, Alan T., and Rex Davis. "Equity in Regional Service Provision." *Journal of Regional Science* 41, no. 4 (2001): 577–600.

Naess, Petter. "'New Urbanism' or Metropolitan-Level Centralization?: A Comparison of the Influences of Metropolitan-Level and Neighborhood-Level Urban Form Characteristics on Travel Behavior." *Journal of Transport and Land Use* 4, no. 1 (2011): 25–44.

National Cooperative Highway Research Program. *Default Values for Highway Capacity and Level of Service Analyses*. NCHRP Report 599. Washington DC: National Cooperative Highway Research Program, 2008.

National Research Council. *Key Transportation Indicators: Summary of a Workshop*. Edited by Janet Norwood and Jamie Casey, Committee on National Statistics, Division of Behavioral and Social Sciences and Education. Washington, DC: National Academy Press, 2002. https://www.nap.edu/read/10404/chapter/420.

National Transit Database. *NTD Annual Database*. https://www.transit.dot.gov/ntd/ntd-data.

Nelles, Jen. "Regionalism Redux: Exploring the Impact of Federal Grants on Mass Public Transit Governance and Political Capacity in Metropolitan Detroit." *Urban Affairs Review* 49, no. 2 (March 1, 2013): 220–53.

Nelson, Arthur C. *Planner's Estimating Guide*. Washington DC: American Planning Association Planners Press, 2004.

Nelson, Arthur C., and Joanna Ganning. *National Study of BRT Development Outcomes*. National Institute for Transportation and Communities Final Report NITC-RR-650. Portland, OR: National Institute for Transportation and Communities, 2015. http://ppms.trec.pdx.edu/media/project_files/NITC-RR-650_ELXlH8p.pdf.

Nussbaum, Martha C. "Capabilities as Fundamental Entitlements: Sen and Social Justice." *Feminist Economics* 9, nos. 2–3 (2003): 33–59.

Nussbaum, Martha C. *Women and Human Development: The Capabilities Approach*. Cambridge: Cambridge University Press, 2000.

O'Toole, Randal. *Gridlock: Why We're Stuck in Traffic and What to Do About It*. Washington, DC: Cato Institute, 2010.

Overgaard, Knud R. *Traffic Estimation in Urban Transportation Planning*. Copenhagen: Danish Academy of Technical Sciences, 1966.

Owen, Andrew, and David M. Levinson. *Access Across America: Transit 2014 Data*. Minneapolis: University of Minnesota Digital Conservancy, 2014. http://dx.doi.org/10.13020/D6MW2Q.

Owen, Andrew, and David M. Levinson. "Modeling the Commute Mode Share of Transit Using Continuous Accessibility to Jobs." *Transportation Research Part A* 74 (2015): 110–22.

Owen, Andrew, Brendan Murphy, and David Levinson. *Access Across America: Auto 2015.* Report CTS 16-07. Minneapolis: Accessibility Observatory, Center for Transportation Studies, Department of Civil, Environmental, and Geo-Engineering, University of Minnesota, 2016. http://access.umn.edu/research/america/auto/2015/.

Owen, Andrew, Brendan Murphy, and David Levinson. *Access Across America: Transit 2015.* Report CTS 16-09. Minneapolis: Accessibility Observatory, Center for Transportation Studies, University of Minnesota, 2016.

Papa, Enrica, and António Ferreira. "'Moving to Access' in Transport Planning: Identifying Barriers, Designing Strategies." Presented at AESOP Congress, Lisbon, July 2017. http://westminsterresearch.wmin.ac.uk/19955/1/AESOP%20Enrica%20Papa%20Antonio%20Ferreira.pdf.

Papa, Enrica, Cecilia Silva, Marco Te Brömmelstroet, and Angela Hull. "Accessibility Instruments for Planning Practice: A Review of European Experiences." *Journal of Transport and Land Use* 3, no. 3 (2015): 1–20.

Paulssen, Marcel, Dirk Temme, Akshay Vij, and Joan L. Walker. "Values, Attitudes and Travel Behavior: A Hierarchical Latent Variable Mixed Logit Model of Travel Mode Choice." *Transportation* 41, no. 4 (2014): 873–88.

Pelham, Thomas G. "Transportation Concurrency, Mobility Fees, and Urban Sprawl in Florida." *Urban Lawyer* 42, no. 4 (2010): 105.

Pendall, Rolf. "Do Land Use Controls Cause Sprawl?" *Environment and Planning B* 26 (1999): 555–71.

Pendall, Rolf. "Local Land Use Regulation and the Chain of Exclusion." *Journal of the American Planning Association* 66, no. 2 (2000): 125–42.

Pérez, Patricio E., Francisco J. Martínez, and Juan de Dios Ortúzar. "Microeconomic Formulation and Estimation of a Residential Location Choice Model: Implications for the Value of Time." *Journal of Regional Science* 43, no. 4 (2003): 771–89.

Polzin, Steven E., Ram M. Pendyala, and Sachin Navari. "Development of Time-of-Day–Based Transit Accessibility Analysis Tool." *Transportation Research Record: Journal of the Transportation Research Board*, no. 1799 (2002): 35–41.

Prashker, Joseph, Yoram Shiftan, and Pazit Hershkovitch-Sarusi. "Residential Choice Location, Gender and the Commute Trip to Work in Tel Aviv." *Journal of Transport Geography* 16, no. 5 (2008): 332–41.

Preston, John, and Fiona Rajé. "Accessibility, Mobility and Transport-Related Social Exclusion." *Journal of Transport Geography* 15, no. 3 (2007): 151–60.

Preston, Valerie, and Sara McLafferty. "Spatial Mismatch Research in the 1990s: Progress and Potential." *Papers in Regional Science* 78 (1999): 387–402.

Proffitt, David G., Keith Bartholomew, Reid Ewing, and Harvey J. Miller. "Accessibility Planning in American Metropolitan Areas: Are We There Yet?" *Urban Studies* 15 (2017): 1–26.

Pucher, John, and John L. Renne. "Socioeconomics of Urban Travel: Evidence from the 2001 NHTS." *Transportation Quarterly* 57, no. 3 (2003): 49–77.

Puget Sound Regional Council. *Transportation 2040 Plan*. Seattle, WA: Puget Sound Regional Council, 2010.

Purvis, Charles. "Data and Analysis Methods for Metropolitan-Level Environmental Justice Assessment." *Transportation Research Record: Journal of the Transportation Research Board*, no. 1756 (2001): 15–21.

Redmond, Lothlorien S., and Patricia L. Mokhtarian. "The Positive Utility of the Commute: Modeling Ideal Commute Time and Relative Desired Commute Amount." *Transportation* 28, no. 2 (2001): 179–205.

Rodrigue, Jean-Paul, Claude Comtois, and Brian Slack. *The Geography of Transport Systems*, New York: Routledge, 2009.

Rollison, Tamara, and Brian Coy. "Gov. McAuliffe Announces Virginia's First Prioritization Process Scores Nearly 300 Proposed Transportation Projects." Virginia Department of Transportation, January 19, 2016. http://www.virginiadot.org/newsroom/statewide/2016/gov._mcauliffe_announces_virginias91759.asp.

Roskie, Jamie Baker, and Janna Blasingame Custer. "Adequate Public Facilities Ordinances: A Comparison of Their Use in Georgia and North Carolina." *Southeastern Environmental Law Journal* 15 (2006): 345–94.

Salomon, Ilan, and Patricia L. Mokhtarian. "What Happens When Mobility-Inclined Market Segments Face Accessibility-Enhancing Policies?" *Transportation Research Part D* 3, no. 3 (1998): 129–40.

Samuelson, Paul A. "The Pure Theory of Public Expenditure." *Review of Economics and Statistics* 36, no. 4 (1954): 387–89.

San Antonio–Bexar County Metropolitan Planning Organization. *Mobility 2040*. San Antonio, TX, 2015.

San Francisco Planning Department. "San Francisco Bicycle Plan Environmental Impact Report." Vol. 1, V.A. 1-1 and V.A. 2-6. 2009.

Sanchez, Thomas W., and Marc Brenman. *The Right to Transportation: Moving to Equity*. Chicago: American Planning Association, 2007.

Sanchez, Thomas W., Qing Shen, and Zhong-Ren Peng. "Transit Mobility, Jobs Access and Low-Income Labour Participation in US Metropolitan Areas." *Urban Studies* 41, no. 7 (2004): 1313–31.

Sanchez, Thomas W., Rich Stolz, and Jacinta S. Ma. *Moving to Equity: Addressing Inequitable Effects of Transportation Policies on Minorities*. Cambridge, MA: Civil Rights Project at Harvard University, 2003.

Schachter, Jason. *Geographical Mobility: March 1999 to March 2000*. Current Population Reports P20-538. Washington, DC: U.S. Census Bureau, 2001.

Scherer, Milena. "Is Light Rail More Attractive to Users Than Bus Transit?: Arguments Based on Cognition and Rational Choice." *Transportation Research Record*, no. 2144 (2010): 11–19.

Schirmer, Patrick M., Michael A. B. van Eggermond, and Kay W. Axhausen. "The Role of Location in Residential Location Choice Models: A Review of Literature." *Journal of Transport and Land Use* 7, no. 2 (2014): 3–21.

Schrank, David, Bill Eisele, Tim Lomax, and Jim Bak. *2015 Urban Mobility Scorecard*. College Station: Texas Transportation Institute, Texas A&M University, 2015.

Schrank, David, and Tim Lomax. *The 2007 Urban Mobility Report*. College Station: Texas Transportation Institute, Texas A&M University, 2007.

Schrank, David, and Tim Lomax. *2009 Urban Mobility Report*. College Station: Texas Transportation Institute, Texas A&M University, 2009.

Schroeder, Bastian J., Christopher M. Cunningham, Daniel J. Findley, Joseph E. Hummer, and Robert S. Foyle. *Manual of Transportation Engineering Studies*. 2nd ed. Washington DC: Institute of Transportation Engineers, 2010.

Schweitzer, Lisa, and Abel Valenzuela Jr. "Environmental Injustice and Transportation: The Claims and the Evidence." *Journal of Planning Literature* 18, no. 4 (2004): 383–98.

Scott, Darren, and Mark Horner. "Examining the Role of Urban Form in Shaping People's Accessibility to Opportunities: An Exploratory Spatial Data Analysis." *Journal of Transport and Land Use* 1, no. 2 (2008): 89–119.

Sen, Amartya. "The Concept of Development." In *Handbook of Development Economics*, edited by Hollis Chenery and Thirukodikaval N. Srinivasen, 9–26. London: Elsevier, 1988.

Sen, Amartya. *Development as Freedom*. New York: Knopf, 1999.

Sen, Amartya. *Inequality Reexamined*. New York: Russell Sage Foundation, 1992.

Sen, Amartya. "Well-Being, Agency and Freedom: The Dewey Lectures 1984." *Journal of Philosophy* 82, no. 4 (1985): 169–221.

Shared-Use Mobility Center. *Shared-Use Mobility Reference Guide*. Chicago: Shared-Use Mobility Center, 2015. http://sharedusemobilitycenter.org/wp-content/uploads/2015/09/SharedUseMobility_ReferenceGuide_09.25.2015.pdf.

Shen, Qing. "Location Characteristics of Inner-City Neighborhoods and Employment Accessibility of Low-Wage Workers." *Environment and Planning B* 25, no. 3 (1998): 345–65.

Social Exclusion Unit. *Making the Connections: Final Report on Transport and Social Exclusion*. London: Office of the Deputy Prime Minister, 2003.

Southeast Michigan Council of Governments. *2030 Regional Transportation Plan for Southeast Michigan*. Detroit: Southeast Michigan Council of Governments, 2004. http://s3.amazonaws.com/zanran_storage/library.semcog.org/Content Pages/2455560602.pdf.

Squires, Gregory D., and Charis E. Kubrin. "Privileged Places: Race, Uneven Development and the Geography of Opportunity in Urban America." *Urban Studies* 42, no. 1 (2005): 47–68.

Srour, Issam, Kara M. Kockelman, and Travis P. Dunn. "Accessibility Indices: Connection to Residential Land Prices and Location Choices." *Transportation Research Record: Journal of the Transportation Research Board*, no. 1805 (2002): 25–34.

Staley, Sam, and Adrian T. Moore. *Mobility First: A New Vision for Transportation in a Globally Competitive Twenty-First Century*. Lanham, MD: Rowman & Littlefield, 2009.

Steinberg, Michael W. "Making Sense of Environmental Justice." *Forum for Applied Research and Public Policy* 15, no. 3 (2000): 82–89.

Stevens, Mark R. "'Does Compact Development Make People Drive Less?': The Answer Is Yes." *Journal of the American Planning Association* 83, no. 1 (2016): 7–18.

Stewart, John Q. "Demographic Gravitation: Evidence and Applications." *Sociometry* 11, no. 1 (1948): 31–58.

Stokenberga, Aiga. "Does Bus Rapid Transit Influence Urban Land Development and Property Values: A Review of the Literature." *Transport Reviews* 34, no. 3 (2014): 276–96.

Stopher, Peter R., and Arnim H. Meyburg. *Urban Transportation Modeling and Planning*. Lexington, MA: Lexington Books, 1975.

Sundquist, Eric. "Virginia DOT's SMART Scale Project Prioritization Process." Paper presented at the *Transportation Research Board 96th Annual Meeting*, Washington, DC, January 8–12, 2017.

Taylor, Brian D., and Paul M. Ong. "Spatial Mismatch or Automobile Mismatch?: An Examination of Race, Residence, and Commuting in U.S. Metropolitan Areas." *Urban Studies* 32, no. 9 (1995): 1537–57.

Taylor, Brian D., and Alexandra Tassiello Norton. "Paying for Transportation: What's a Fair Price?" *Journal of Planning Literature* 24, no. 1 (2009): 22–36.

Te Brömmelstroet, Marco, and Luca Bertolini. "Integrating Land Use and Transport Knowledge in Strategy-Making." *Transportation* 37, no. 1 (2010): 85–104.

Thomas, Trevor, Andrew Mondschein, Taner Osman, and Brian D. Taylor. "Not So Fast?: Examining Neighborhood-Level Effects of Traffic Congestion on Job Access. *Transportation Research Part A* 113 (2018): 529–41.

Thorsen, Inge, and Jens P. Gitlesen. "A Simulation Approach to Studying the Sensitivity of Commuting-Flow Predictions with Respect to Specific Changes in Spatial Structure." *Environment and Planning A* 34, no. 2 (2002): 271–88.

Tomer, Adie, Elizabeth Kneebone, Robert Puentes, and Alan Berube. *Missed Opportunity: Transit and Jobs in Metropolitan America*. Washington, DC: Brookings Institution, 2011.

Townsend, Peter. *Poverty in the United Kingdom: A Survey of Household Resources and Standards of Living*. Berkeley: University of California Press, 1979.

Transit Cooperative Research Program. *Civil Rights Implications of the Allocation of Funds between Bus and Rail*. Legal Research Digest 27. Washington, DC: Transportation Research Board, 2008.

Transit Cooperative Research Program. *Transit Capacity and Quality of Service Manual*. 2nd ed. Washington, DC: Transportation Research Board, 2003.

Transit Cooperative Research Program. *Transit Capacity and Quality of Service Manual*. 3rd ed. Washington, DC: Transportation Research Board, 2013.

"Transportation Concurrency." Chap. 12.24 of the City of Snoqualmie (Washington) Municipal Code. http://www.codepublishing.com/WA/Snoqualmie/.

Transportation Research Board. *Driving and the Built Environment: The Effects of Compact Development on Motorized Travel, Energy Use, and CO_2 Emissions*. Washington, DC: National Academies Press, 2009.

Transportation Research Board. *Expanding Metropolitan Highways: Implications for Air Quality and Energy Use*. Special Report 245. Washington, DC: National Academies Press, 1995.

Transportation Research Board. *Highway Capacity Manual*. Washington, DC: National Academies Press, 1994.

Transportation Research Board. *Highway Capacity Manual 2010*. http://hcm.trb.org/?qr=1.

Transportation Research Board. *Highway Capacity Manual: A Guide for Multimodal Mobility Analysis*. 6th ed. Washington, DC: Transportation Research Board, 2016.

Transportation Research Board. *Travel Estimation Techniques for Urban Planning*. NCHRP Report 365. Washington, DC: National Academies Press, 1998.

Tumlin, Jeffrey. *Sustainable Transportation Planning: Tools for Creating Vibrant, Healthy, and Resilient Communities*. Hoboken, NJ: Wiley, 2012.

Turner, Christopher G. "A Model Framework for Transportation and Community Plan Analysis." *Journal of the American Planning Association* 38, no. 5 (1972): 325–31.

United States Census Bureau. *American Community Survey*. U.S. Census Bureau's American Community Survey Office, 2006–2010.

United States Census Bureau. *2013 American Community Survey*. U.S. Census Bureau's American Community Survey Office. http://factfinder2.census.gov.

United States Census Bureau. *LODES Data*. Longitudinal-Employer Household Dynamics Program, 2015. http://lehd.ces.census.gov/data/lodes/.

United States Code of Federal Regulations. Title 23, Highways. 2006.

United States Department of Labor. *Economic News Release: Employee Tenure Summary*. Washington, DC: Bureau of Labor Statistics, 2010. http://www.bls.gov.proxy.lib.umich.edu/news.release/tenure.nr0.htm.

United States Department of Transportation. *Highway Statistics 2002*. Washington, DC: Bureau of Highway Statistics, 2002.

United States Department of Transportation. *2001 National Household Travel Survey*. Washington, DC: Federal Highway Administration and Bureau of Transportation Statistics, 2003.

United States Department of Transportation. *2009 National Household Travel Survey*. Washington, DC: Federal Highway Administration, 2011.

United States General Accounting Office. *Bus Rapid Transit Shows Promise*. Washington DC: U.S. General Accounting Office, 2001.

van Wee, Bert. "Evaluating the Impact of Land Use on Travel Behavior: The Environment versus Accessibility." *Journal of Transport Geography* 19, no. 6 (2011): 1530–33.

van Wee, Bert, and Karst T. Geurs. "Discussing Equity and Social Exclusion in Accessibility Evaluations." *European Journal of Transport and Infrastructure Research* 11, no. 4 (2011): 350–67.

van Wee, Bert, and Karst T. Geurs. "The Role of Accessibility in Urban and Transport Planning." In *Handbook on Transport and Urban Planning in the Developed World*, edited by Michiel C. J. Bleimer, Corinne Mulley, and Claudine J. Moutou, 53–66. Cheltenham, UK: Edward Elgar, 2016.

Vega, Amaya, and Aisling Reynolds-Feighan. "A Methodological Framework for the Study of Residential Location and Travel-to-Work Mode Choice under Central and Suburban Employment Destination Patterns." *Transportation Research Part A* 43, no. 4 (2009): 401–19.

Virginia Department of Transportation. *Smart Scale Technical Guide*. Richmond: Virginia Department of Transportation, 2016.

Wachs, Martin. "Learning from Los Angeles: Transport, Urban Form, and Air Quality." *Transportation* 20, no. 4 (1993): 329–54.

Wachs, Martin, and T. Gordon Kumagai. "Physical Accessibility as a Social Indicator." *Socio-Economic Planning Sciences* 7, no. 5 (1973): 437–56.

Wachs, Martin, and Brian D. Taylor. "Can Transportation Strategies Help Meet the Welfare Challenge?" *Journal of the American Planning Association* 64, no. 1 (1998): 15–19.

Waddell, Paul, Chandra Bhat, Naveen Eluru, Liming Wang, and Ram Pendyala. "Modeling Interdependence in Household Residence and Workplace Choices." *Transportation Research Record: Journal of the Transportation Research Board*, no. 2003 (2007): 84–92.

Walker, Jarrett. *Human Transit: How Clearer Thinking about Public Transit Can Enrich Our Communities and Our Lives*. Washington, DC: Island Press, 2012.

Walton, William. "Linking New Development to Traffic Generation: The Deficiencies of the Current Methodology of Traffic Impact Analysis." *Town Planning Review* 69, no. 4 (1998): 375–96.

Weiner, Edward. *Urban Transportation Planning in the United States: A Historic Overview*. Westport, CT: Praeger, 1999.

Widener, Michael J., Steven Farber, Tijs Neutens, and Mark W. Horner. "Using Urban Commuting Data to Calculate a Spatiotemporal Accessibility Measure for Food Environment Studies." *Health & Place*, no. 21 (2013): 1–9.

Wilson, Alan G. "A Family of Spatial Interaction Models, and Associated Developments." *Environment and Planning A* 3, no. 1 (1971): 1–32.

Wilson and Company. "Interchange Access Change Request." Unpublished memo. Wilson and Company, 2013.

Wyly, Elvin K., and Daniel J. Hammel. "Gentrification, Segregation, and Discrimination in the American Urban System." *Environment and Planning A* 36, no. 7 (2004): 1215–41.

Zhang, Ming. "Bus versus Rail: Meta-Analysis of Cost Characteristics, Carrying Capacities, and Land Use Impacts." *Transportation Research Record: Journal of the Transportation Research Board*, no. 2110 (2009): 87–95.

Zondag, Barry, and Marits Pieters. "Influence of Accessibility on Residential Location Choice." *Transportation Research Record: Journal of the Transportation Research Board*, no. 1902 (2005): 63–70.

Index

The letter *f* following a page number indicates a figure; *t* indicates a table.

CPSIA information can be obtained
at www.ICGtesting.com
Printed in the USA
FSHW012243280919
62472FS

9 781501 716089